The Biology of Investing

Why do people's financial and economic preferences vary so widely? "Nurture" variables such as socioeconomic factors partially explain these differences, but scientists have been discovering that "nature" also plays an important role. This is the first book to bring together these scientific insights for a holistic view of the role of human biology in financial decision making.

Geneticists are now examining which genetic markers are associated with financial and economic preferences. Neuroscientists are determining where in the brain financial decisions are made and how that varies between people. Endocrinologists relate the level of different hormones circulating in the body to financial risk taking. Researchers are exploring how physiology and environmental conditions influence investment decisions, and how three types of cognitive ability play essential roles in investment success. The exciting and relevant work being done in these academic silos has generally not been transmitted among the scientific areas, or to industry. For the first time, this book integrates all these areas, explaining the myriad ways in which a person's biology influences their investing decisions.

Financial analysts, advisors, market participants, and upper-level undergraduate and postgraduate students of behavioral finance, behavioral economics, and investing will find this book invaluable, enabling a deeper understanding of investors' decision-making processes.

To further ensure that this new material is accessible to students, PowerPoint slides are available online for instructors' use.

John R. Nofsinger is the William H. Seward Chair in International Finance at the University of Alaska Anchorage and is a portfolio manager at Denali Advisors, LLC.

Corey A. Shank is an Assistant Professor of Finance at Dalton State College.

The Biology of Investing
Nature, Nurture, Physiology, & Cognition

John R. Nofsinger and Corey A. Shank

Routledge
Taylor & Francis Group

NEW YORK AND LONDON

First published 2020
by Routledge
52 Vanderbilt Avenue, New York, NY 10017

and by Routledge
2 Park Square, Milton Park, Abingdon, Oxon OX14 4RN

Routledge is an imprint of the Taylor & Francis Group, an informa business

Visit the eResources: www.routledge.com/9780367443399

Library of Congress Cataloging-in-Publication Data
Names: Nofsinger, John R., author. | Shank, Corey A., author.
Title: The biology of investing : nature, nurture, physiology, & cognition / John R. Nofsinger and Corey A. Shank.
Description: 1 Edition. | New York : Routledge, 2020. | Includes bibliographical references and index.
Identifiers: LCCN 2019055692 | ISBN 9780367444143 (hardback) | ISBN 9780367443399 (paperback) | ISBN 9781003009566 (ebook)
Subjects: LCSH: Investments–Psychological aspects. | Decision making–Psychological aspects. | Risk-taking (Psychology) | Human biology.
Classification: LCC HG4515.15 .N6428 2020 | DDC 332.601/9–dc23
LC record available at https://lccn.loc.gov/2019055692

ISBN: 978-0-367-44414-3 (hbk)
ISBN: 978-0-367-44339-9 (pbk)
ISBN: 978-1-003-00956-6 (ebk)

Typeset in Sabon
by Taylor & Francis Books

I dedicate this book to my wife and best friend, Anna.

John

To my wife Ashley, marrying you is a dream come true.
To my parents Amy and Bobby, who I quite literally owe my life and so much more for their sacrifice.

Corey

We are also both grateful for manuscript reviews by Hannah Frenier and Amy Shank.

Contents

List of illustrations   xiii

1   Biology and Psychology in Finance   1

   Nature versus Nurture   2
      What Can We Learn from the Financial Decisions
         of Twins?   3
      Investment of Adoptees and the Human Genome   4
      Do Men and Women Invest Differently?   4
   Physiology   5
      Brain Function and Financial Decisions   5
      The Influence of Hormones on Financial Risk Taking   6
      Sleep, Coffee, and Investing   6
      How Wellness Influences Financial Decisions   6
   Cognitive Outcomes   7
      The Emotional and Moody Investor   7
      Environmental Factors in Financial Decision Making   8
      Personality and Investing   8
      Intelligence and Investment Performance   9
      Cognitive Aging and Diminished Decision Making   9
   A Possible Future   10

SECTION I
Nature versus Nurture   11

2   Genetics: Twin Investment Behavior   13

   Twin Research: Nature, Nurture, and Unique Experiences   13
   Saving and Homeownership   14
      Saving   14

Homeownership   15
Risk Attitudes   16
Investing   18
  Stock Market Participation   18
  Pension Choices   20
  Investing Style: Value versus Growth   21
Behavioral Biases   22
Well-being: Giving and Happiness   24
  Giving   24
  Happiness   25
Summary   26
Questions   26

3 More Genetics: Investments of Adoptees and the
   Human Genome   28

  Adoption   28
    Introduction to Adoption Studies: Education and Income 29
    Investing Decisions   32
  Genome   34
    The Genome-wide Association Method   35
    Investing   36
    Risk Aversion   38
  Summary   39
  Questions   40

4 Do Men and Women Invest Differently?   42

  Determinants of Gender Differences   42
    Gender Biology   43
    Societal Differences   43
  Financial Experiment Research   44
  Retirement Accounts   45
  Trading Behavior   46
  Gender in Corporate Finance   49
  Nature or Social Norms?   50
    Feminism and Masculinity   50
    Financial Literacy   52
  The Gender of Others   54
  Summary   56
  Questions   56

SECTION II
Physiology   59

5   Brain Function and Financial Decisions   61

   The Brain   61
      Anatomy of the Brain   61
      Neurotransmitters   64
      Types of Decision Processes   64
   Economic Decisions and the Brain   65
      Financial Risk and Reward   65
      Evidence from Brain Damage   68
      Biological Foundations of Behavioral Finance   69
   Genoeconomics   73
      Genoeconomics Research   73
      Genes and Receptors   73
      DRD4 and Financial Decision Making   74
      MAOA and Financial Decision Making   75
   Summary   76
   Questions   76

6   The Influence of Hormones on Financial Risk Taking   79

   Hormones   79
      Hormonal Physiology   79
   Circulating Testosterone and Cortisol Levels   81
      Testosterone and Economic Risk Taking   81
      Cortisol and Economic Risk Taking   82
      The Dual Hormone Hypothesis   82
      Testosterone, Cortisol, and Financial Decisions   83
   Testosterone Proxies   88
      2D:4D Ratio and Facial Masculinity   88
      Testosterone Proxies and Economic Risk Taking   89
      Testosterone Proxies and Financial Decision Making   89
      Testosterone Proxy Validity   91
   Summary   92
   Questions   92

7   Sleep, Coffee, and Investing   95

   Sleep Physiology   95
   Sleep and Economic Risk Taking   96
   Sleep, Economic Risk Taking, and Brain Activation   98

Sleep and Financial Decisions	99
Country-wide Analysis	99
Individual Focused Experiments	100
Sleep Deprivation, Stimulants and Risk Taking	101
Summary	104
Questions	104

8 How Wellness Influences Financial Decisions — 107

Health and Obesity	107
Exercise and Diet	108
Physical Health and Financial Decisions	109
Health and Portfolio Choice	109
Health and Risk Taking: Experimental Evidence	110
Employee Health and Company Performance	111
Society Illness and Stock Market Performance	112
Mental Health and Financial Decisions	113
Impact of Finance on Health	114
Summary	116
Questions	117

SECTION III
Cognitive Outcomes — 121

9 The Emotional and Moody Investor — 123

Emotions, Mood, and Physiology	123
Emotions and Economic Decisions	124
Physiological Response to Economic Decisions	124
Emotions, Mood, and Experimental Finance	124
Loss Aversion and Affect	129
Affect and Ethical Decision Making	132
Aggregate Mood and the Stock Market	132
Summary	134
Questions	135

10 How Environmental Factors Impact Financial Decisions — 137

Weather, Mood, and Financial Decisions	137
Weather and Mood	137
Weather and Experimental Finance	138
Weather and Stock Markets	138

Seasonal Affective Disorder   138
   Seasonal Affective Disorder and Physiology   138
   Seasonal Affective Disorder and Experimental Finance   139
   Seasonal Affective Disorder and Stock Markets   139
   Seasonal Affective Disorder and Financial Analysts   141
Air Pollution, Cognition, and Financial Decisions   142
   Air Pollution and Cognition   142
   Air Pollution and the Stock Market   143
Allergies and Financial Decisions   145
Natural Disasters and Financial Decision Making   145
   Natural Disasters and Behavior   145
   Natural Disasters and Financial Decisions   146
Terrorism and the Stock Market   147
   Terrorism and Behavior   147
   Terrorism and the Stock Market   148
Summary   148
Questions   148

11 The Personality of a Successful Investor   152

Big Five Personality Traits and Financial Decisions   152
   Big Five Personality Traits   152
   Big Five Personality Traits and Neural Activity   153
   Big Five Personality Traits and Experimental Risk Taking   154
   Big Five Personality Traits and Experimental Investments   155
   Big Five Personality Traits and Investments   156
   Big Five Personality Traits and Corporate Finance   157
   Big Five Personality Traits and Personal Finances   158
The Dark Triad and Financial Decisions   159
   The Dark Triad   159
   The Dark Triad and Business   159
   The Underlying Conditions of the Dark Triad   160
   The Dark Triad and Financial Decisions   160
Noncognitive Abilities and Financial Decisions   163
   Noncognitive Abilities   163
   Noncognitive Abilities and Business   163
   Overconfidence, Optimism, and Financial Decisions   163
   Noncognitive Abilities and Financial Decisions   165
Summary   165
Questions   166

12   Types of Intelligence and Investment Performance   169

  Types of Cognitive Processing   169
  Cognitive Processing and Investing   170
    IQ and Investing Decisions   170
    Analytical versus Intuitive Thinking and
      Investing Decisions   173
    Measuring Analytical and Intuitive Tendencies:
      Cognitive Reflection   174
    Cognitive Reflection and Behavioral Biases   175
    The Theory of Mind and Investing Decisions   178
    How Each Type of Intelligence Impacts
      Investing Decisions   179
  Summary   180
  Questions   181

13   Impaired Cognitive Function and Diminished Decision Making   183

  Aging and Cognitive Ability   183
    Decision Processing and Aging   184
  Aging and Financial Literacy   185
    Financial Knowledge   185
    Seeking Financial Advice   187
    Credit Choices   188
  Aging and Financial Fraud   189
    Cognitive Aging and Financial Fraud   190
  Investment Decisions of Seniors   191
    Aging and Financial Risk Aversion   191
    Aging and Portfolio Construction and Performance   193
    Aging and Behavioral Biases   194
    Improving Decisions of Older Adults   196
    Use It or Lose It   196
  Summary   197
  Questions   197

  Index   200

Illustrations

Figures

2.1   Decision Correlation between Groups of Twins   19
2.2   Portion of the Psychological Bias Explained by Genetics   23
3.1   Variance Decomposition Attributable to Nature, Nurture, and Unique Experiences   30
4.1   Gender and Mutual Fund Choice   48
4.2   Risk Aversion Scores of Parents   55
5.1   The Main Areas of the Brain   62
5.2   The Limbic System   63
6.1   Testosterone and Traders' Profits and Losses   83
6.2   Cortisol Levels by Standard Deviation of Profit and Loss   84
6.3   Testosterone, Cortisol, and Excess Return   85
6.4   Cortisol, Testosterone, and Investment Biases in a Financial Trading Simulation   86
6.5   Cortisol and Testosterone and Risk Taking in a Double Auction Trading Game   87
6.6   Alpha by Testosterone Exposure   90
7.1   Sleep Deprivation, Stimulants, and the BART   102
7.2   Sleep Deprivation, Stimulants, and the Iowa Gambling Task   103
8.1   Prevalence of Obesity in the United States   108
8.2   Health and Asset Holdings   109
9.1   Affective Cues and the Endowment Effect   125
9.2   Affective States and Risk Taking in Gain/Loss Decision Frames   128
9.3   Rebalancing Frequency and Bond Fund Allocation   130
9.4   Average Amount Bet Based on Treatment and Experience   131
10.1   Daily Stock Return and Air Quality   143
10.2   Underperformance of Stocks Based on Air Pollution on Date of Purchase   144
12.1   IQ and Stock Market Participation   171
13.1   Financial Cognitive Ability and Age   184
13.2   Financial Knowledge and Confidence   186

13.3   Annualized Abnormal Return by Age Group   193
13.4   Percent Selecting Optimal or Nearly Optimal Group by Age   195

Tables

3.1   Sources of Korean Norwegian Adoptees' Decisions   33
5.1   Functions and Locations in the Brain   63
11.1 Characteristics of the Big Five Personality Traits   153

1 Biology and Psychology in Finance

The evolution of financial theory has moved at a blistering pace in the past few decades. The traditional standard finance approach is based on classical decision theory, which assumes that people and markets are completely rational. The underlying assumptions are that people have access to perfect information, which is costless the moment it is released, can apply unlimited brain processing resources, have internally consistent preferences, and use expected utility theory to maximize the benefits of the decision. In other words, when money is involved, people make unbiased decisions to maximize their own self-interest, and everyone should come to the same conclusions about risk and return. Furthermore, any error a person makes is not predictable and not correlated with other investors' decisions so that market prices are not impacted. Unfortunately, these assumptions are not valid as people frequently make predictable irrational decisions, especially when money is involved. In fact, even Alfred Marshall, the economist who developed the supply and demand curves, recognized that people are affected by physiological influences when he said, "Economics is a branch of biology broadly interpreted."

Standard finance theory has been challenged by behavioral finance as decades of research have shown that people do not meet the rationality benchmark of standard finance. Indeed, people frequently make the same predictable errors that others are making.[1] The ideal of the rational investor has been replaced with that of the normal investor.[2] While standard finance dictates how investors should behave when making financial and economic decisions to maximize their wealth, behavioral finance describes how they actually behave. Behavioral finance tries to explain actual behavior through psychological, cognitive, and emotional factors that often lead a person to biased, non-rational, financial decisions. As such, behavioral finance and behavioral economics have become mainstream financial paradigms. Some of the early pioneers in behavioral finance have received the Nobel Memorial Prize in Economic Sciences. They include Daniel Kahneman and Vernon L. Smith in 2002, Robert J. Shiller in 2013, and Richard Thaler in 2017.

When people make consistent errors, psychologists call them cognitive biases. One common cognitive bias is the bandwagon effect (also known as groupthink or herd behavior), whereby people tend to believe in things or ideas simply because the vast majority of others do. Similarly, consider the endowment effect, which is the tendency for people to put a higher value on objects that they own compared to a similar object that someone else owns. One aspect of the endowment effect is the IKEA effect, whereby people place a higher value on items, such as furniture, that they assembled themselves.

Behavioral finance calls these cognitive biases investment biases when describing investors' decisions about their investments. For example, the literature consistently shows that investors who commit these investment biases earn lower returns compared to those who do not. However, if investors were rational, they would not make these financial decisions. As such, behavioral finance describes investors' decision making much better than the standard finance ideal over recent decades. Furthermore, there is a good deal of variation in the decisions which people make that has not been explained. Why do some people succumb to a particular investment bias while others do not? Why does one investor make different decisions at two different times when faced with the same situation? Although behavioral finance has worked hard to discover these investment biases, it has been silent about the sources of the biases. What is the source of some people using an analytical mode of thinking while others are driven to use an intuitive approach? Behavioral finance describes people as being loss averse, but what drives some people to be more loss averse than others? One answer is described in this book – biology.

Biology can influence behavior in a transient, quasi-permanent, and permanent manner. Temporary impacts on decision making include the biological ramifications that are transitory, such as sleep deprivation, circulating hormone levels (such as testosterone), emotions, moods, or the environment (such as the weather). For example, people are likely to make a different financial decision when they are sleep deprived compared to when they are well rested. Tired investors are temporarily more risk and loss averse. Health status also impacts financial decision making as people in poor health tend to take less risk than healthy people. However, people tend to stay healthy or unhealthy for long periods of time, and these factors do not fluctuate as do sleep deprivation or mood. Thus, the physical and mental health status of a person tends to be quasi-permanent and can bias economic choices for many years. Finally, many biological influences are permanent. A person's genetic code (i.e., their DNA or gender) might make them more likely to succumb to a specific bias, yet this genetic code cannot be changed.

The remainder of this chapter sets up the biology topics of the book. The topics are divided into three sectional themes: nature versus nurture; physiology; and cognitive outcomes.

Nature versus Nurture

To what degree are human behaviors innate versus learned? Is a newborn baby's psychology a blank slate, or is the child partially pre-programmed from inherited DNA? Are decisions programmed by genome characteristics, or do they stem from a history of experiences? These questions have been studied for centuries. Physical traits such as eye color are determined by specific genes (i.e., nature). Some behavioral characteristics such as temperament may be more influenced by a person's experiences when growing up, (i.e., nurture). More recently, the nature versus nurture debate has included financial risk aversion and investment decision making.

One interesting group of studies examines the behavior of twins. How do different types of twins invest? Note that identical twins share all of their DNA, while fraternal twins share about half of that DNA. Using a large sample of twin financial decisions, researchers can allocate the sources driving the decisions among genes (nature), rearing environment (nurture), and unique adult experiences. Another line of research uses this same idea but uses the decisions of adopted children, biological children, and biological parents. Twin and adoptee studies use birth status as a proxy for genetics. Can we examine a person's genome directly? Yes we can. However, the vast number of genetic variations creates some analysis problems. Finally, there is one obvious chromosome difference that can easily be identified: males and females. There is a plethora of research on the differences in financial decisions made by men and women. However, there are often significant social norms about male and female behavior that can drive a person's decisions. In the end, it appears that behavior is influenced by both nature *and* nurture. Still, the degree to which source has the greatest influence continues to be studied.

What Can We Learn from the Financial Decisions of Twins?

Chapter 2 reviews the studies and findings of the financial decisions of twins. Twin siblings share some portion of DNA and most are raised together. A striking result is that genetically identical twins make many more similar decisions than do fraternal twins. Examples of financial decisions include their preferences for saving, houses, financial risk taking, behavioral biases, and charitable giving. The propensity toward making a specific decision can be portioned between nature (genetics), nurture (shared rearing experience), and unique adult experiences. Depending on the specific decision, nature (i.e., genetics) is found to explain one-fifth to one-half of the variation in preferences between twins. A person's unique experiences explain the largest variation in decision making. The nurture environment in which the twins were raised appears to drive a surprisingly small portion of these decisions. However, the nurture environment has a more significant impact on decisions made by younger adults. The ability for nurture to drive financial decisions declines as a person ages while the contribution of unique experiences rises over time, and the influence of genetic coding persists throughout life.

Investment of Adoptees and the Human Genome

Chapter 3 expands the research line of studying genetics in twin behavior by exploring the financial preferences of adoptees, their parents (biological and adoptive), and adoptive siblings (the biological children of the adoptee's new parents). Adoptees share none of the genetic markers with their adoptive siblings, yet they share a similar nurture (rearing) environment. These studies report that biology plays an essential role in explaining economic and financial outcomes. They report that the nature contribution is about half of the post-birth effects' contribution (i.e., nurture and unique experiences combined for about two-thirds). Thus, genetics explains about one-third of the variation of economic outcomes among people. The studies also report that there is a role model aspect of nurture. Specifically, the economic outcomes of mothers have more influence on the outcomes of daughters, while the outcomes of fathers have more influence on sons.

Adoptee and twin behavior studies use birth status to proxy for genetic coding. Now that the human genome has been mapped, people are exploring their DNA to learn about their cultural history and health vulnerabilities. However, assigning specific genetic markers to behavior is proving to be complicated. The problem is the vast number of genetic markers that differ among people. Trying to locate specific markers that drive financial behavior among the millions of genetic differences using samples of people in the thousands, even hundreds of thousands, often leads to spurious (untrue) results. Nevertheless, social scientists are developing methods to reduce the complexity by grouping the DNA differences among people. A person's specific DNA combination of associated groups is called their polygenic score. Through this method, scientists have discovered the groups of markers associated with educational attainment. That same grouping influences economic outcomes and investment characteristics, like risk aversion, stock market participation, and probabilistic thinking. The polygenic score is the first step in determining the mechanisms for how genetic markers may influence financial decisions.

Do Men and Women Invest Differently?

Popular culture is full of expressions that indicate different expectations, biases, and behaviors between men and women. Some examples are "man up" and "throw like a girl." Assertive men are often considered leaders, while assertive women are labeled bossy. Or consider John Grey's book title *Men Are from Mars, Women Are from Venus.*[3] Phrases like these express social norms or perceptions about the behavior and actions of men and women, whether real or not. Of concern here is whether there are any real differences in financial decisions. If differences exist, are they due to biology (sex) or to social norms (gender)? Chapter 4 first reviews the literature and concludes that men and women exhibit significant differences in financial decisions. Most importantly, there is strong evidence that women take less risk than men, including financial risk. Men tend to exhibit more overconfidence and are more likely to exhibit sensation-seeking behavior which may be causing this risk-taking behavior. Complicating the matter is that differences in financial behavior could be caused by differences in emotional reaction, competitive behavior, or financial literacy. Further exploration illustrates that social norms play an important role in shaping a person's risk preferences. Overall, it appears that some differences between men and women are a detriment to women, while others are beneficial.

Physiology

Human physiology is the biochemical, physical, and mechanical function of the human body. This includes how the brain functions, the level of biochemicals like circulating hormones, and the body's health. Psychologists have been studying how various aspects of physiology impact behavior. It is not likely a surprise that a traumatic brain injury may cause a person to exhibit poor judgment and impulsive behavior. However, even small variations in the function of healthy brains have an impact on behavior. One way the brain stays healthy is to regenerate during sleep. Thus, a significant lack of sleep affects the brain's functions and increases negative emotions like fatigue, anxiety, and depression. As such, a tired person makes different decisions than when well rested. Finally, health also influences a person's behavior. Furthermore, sick or unhealthy people have different economic and risk preferences than healthy people who have good diets and who exercise.

How various aspects of physiology impact financial decisions and investment preferences are explored in the physiology section of the book. For organizational purposes, physiology topics are centered on brain function, hormones, sleep, and wellness.

Brain Function and Financial Decisions

Everybody's brain functions a little differently. The study of the structure and function of the brain during financial and economic decision making is called neuroeconomics. Using instruments like functional magnetic resonance imaging (fMRI) machines, researchers can map the location in the brain that activates when making different decisions. Activities mapped to the frontal lobe show more rational, controlled processes. Conversely, the amygdala activates during emotional and fear responses and is linked to irrational and impulsive decisions. Chapter 5 reviews the literature that compares the location of brain activation and the quality of financial decisions. People who use their amygdala during financial decisions are more likely to make irrational decisions, such as avoiding ambiguous choices and succumbing to loss aversion. Also, risk-reward decisions are mediated by the nucleus accumbens, which is part of the dopaminergic pathway. Finally, genetic variations seem to drive how the brain processes dopamine and serotonin. The presence of these neurotransmitters impacts the amount of financial risk a person takes.

The Influence of Hormones on Financial Risk Taking

Testosterone boosts aggression, competitiveness, and self-esteem, and is known as the male hormone because men have higher levels than women due to biological differences. Additionally, cortisol is known as the stress hormone because it appears in higher levels during periods of high stress. These hormones directly influence the chemical makeup of the brain by altering how neurons send and receive signals. Chapter 6 illustrates how hormones impact financial risk taking. Hormone levels can affect the brain at two different times. First, the amount of testosterone in utero influences the baby's brain and body formation, which influence behavior through adulthood. Second, the level of hormones circulating in the body during the time of a decision affects the decision. Both circulating levels of testosterone and proxies for testosterone in utero are shown to be positively related to risk taking in economic and financial decisions. In addition, cortisol is related to risk aversion in economic and financial situations. Interestingly, higher levels of both testosterone and cortisol are related to irrational financial decisions.

Sleep, Coffee, and Investing

Sleep deprivation impairs everything a person might do, from physical activity to complex decision making. Poor sleep quality directly influences brain functions through decreased activation and neural function. Chapter 7 illustrates that people who are sleep deprived are both risk averse and loss averse. That is, they become less likely to take financial risks than normal, even when those risks offer high returns. Additionally, sleep-deprived investors who are more loss averse than usual are predisposed to irrational financial decisions. Maybe the sleep deprived just need a cup of coffee? Actually, caffeine does not reverse the impact of sleep deprivation. Stimulants make one feel more awake, but do not improve financial decisions.

How Wellness Influences Financial Decisions

It comes as no surprise that people with high wealth can afford the best wellness programs which make health and wealth positively related. However, health status also influences financial choices. For example, Chapter 8 documents that people in poor health are more risk averse. This may be a rational response to expectations about needing their money to cover future medical expenditures, thereby causing them to avoid higher-risk investments. Low amounts of exercise and a poor diet are measures of poor health and are found to be associated with poor investment decisions. Aerobic exercise and improved diet have direct positive effects on cognition and financial decision making. Mental health is also essential, but it seems to impact men and women differently in regards to financial decisions. The health of people in the community also impacts the employers in the region, which can influence how corporations behave. Finally, the economy impacts peoples' health. For example, economic downturns are associated with lower mental well-being and less healthy choices.

Cognitive Outcomes

The brain's acquisition of knowledge and understanding comes through thoughts, experiences, and the senses. One result of cognitive processes is a decision. The acquisition of knowledge and the resulting decision outcomes of cognition are filtered by emotions, moods, and personality. The impact of emotions and moods on investment choices have been studied in the context of behavioral finance. There are different ways to categorize personality. One popular method is known as the "big five" personality traits: extroversion, agreeableness, neuroticism, openness, and conscientiousness. Researchers have found that fundamental differences in personality are related to the neuron volume in different regions of the brain. Thus, personality may literally be wired into the brain. Furthermore, do the more intelligent make better financial decisions? Researchers examine this question by looking at three different types of intelligence: fluid intelligence; cognitive reflection; and theory of mind. Intelligence quotient (IQ) is a measure of fluid intelligence that measures reasoning ability. Cognitive reflection is the tendency to use an analytical mode of thinking rather than an intuitive mode. The theory of mind refers to a person's ability to infer the intentions of others. Finally, how does age influence cognitive ability? As we age, we gain experience, which helps us to make smarter decisions. However, as we age our cognitive abilities start to decline. As such, it is crucial to examine the impact of cognitive aging on financial decisions.

The Emotional and Moody Investor

Traditional theories of finance assume that people act rationally in their own best interests when money is involved. However, that description does not describe the majority of people. Emotions and moods influence people's behavior. Chapter 9 demonstrates that emotion and mood influence peoples' risk taking and financial decision making. When assessing investment opportunities, happier people are more optimistic and more willing to take risks. Less happy people are more critical and less willing to take financial risks. Emotions and moods can aggregate across society to impact the stock market. For example, researchers can measure how worried investors may be about the stock market and find that the stock market tends to fall when investors are more nervous. While emotions impact investment choices, conversely investment outcomes also affect emotions. For example, more frequent checking of one's portfolios fosters increased emotional arousal to losses. This emotional response to gains and losses is related to increased irrational behavior. The mood of a company's chief executive officer (CEO) can also be inferred, and this influences how investors view the stock. Overall, research indicates that mood and emotion can have large impacts on financial decision making and stock market movements.

Environmental Factors in Financial Decision Making

External factors can impact how people view the investment process, especially when those factors influence their mood and health. Chapter 10 reviews how financial decisions are influenced by several environmental factors such as the weather, air pollution, and terrorism. First, sunshine leads to positive moods that cause the stock market to rise, people to take more risks, and investors to be more optimistic. Alternatively, seasonal affective disorder, the negative mood brought on by the decreasing amount of daylight in the fall and winter months, causes less risk taking and more pessimistic views. Air pollution adversely effects cognitive function and mood and thus influences society's investors, which leads to investors making poor financial decisions when pollution is high. Even allergens like pollen cause cognitive problems for investors. Finally, extreme events like natural disasters and terrorist attacks cause investors to be fearful, thus causing more risk-averse investing behavior.

Personality and Investing

Is there a personality for investing? It appears that some personality traits and other noncognitive abilities are better suited to the investment process than others. One crucial aspect is risk aversion. If a person's risk aversion is too high, then it is unlikely that they will purchase assets that will produce sufficiently high returns, which would cause a lower level of wealth over the course of his lifetime and a lower amount of money available for retirement. Chapter 11 examines the impact of personality traits and other noncognitive abilities on financial decisions. The two personality traits of the "big five" that are most frequently related to financial decisions are extroversion, which refers to being outgoing with a positive attitude, and neuroticism, which is characterized by anxiety and depression. Medical researchers find that extroverts tend to have higher levels of dopamine making them happier and more outgoing. Conversely, neurotic individuals tend to have higher levels of norepinephrine and cortisol causing them to have higher levels of anxiety. Overall, the literature shows consistent evidence that extroverted individuals are more willing to take risks while people characterized as neurotic are more likely to be risk averse.

There is a difference between which personality types make better investors and which personality types are attracted to the finance industry and business in general. For example, overconfidence and optimism drive people to take the chances that can lead a person into top management. However, those traits can also cause those CEOs to make irrational decisions because they overestimate their abilities. Ever think that your boss is a narcissistic psychopath? Your thoughts may be justified because research finds that psychopaths are four times more likely to hold positions of power, such as executives in business. These psychopathic traits often lead to individuals getting promotions and moving up in the company due to their ruthlessness to succeed at any cost. However, psychopathy and the other "dark triad" personality traits (Machiavellianism and narcissism) typically have negative consequences for firm performance.

Intelligence and Investment Performance

Chapter 12 examines how fluid intelligence, cognitive reflection, and the theory of mind impact financial decisions. Fluid intelligence is usually measured through IQ standardized tests. Investors with a higher IQ are more likely to invest in the stock market and to hold superior portfolios as measured in several different ways. Conversely, lower IQ investors are more prone to irrational biases and financial decisions. When examining cognitive reflection, evidence shows that investors with an analytical mode of thinking perform better when dealing with the complexity of uncertainty and other financial decisions, while the intuitive thinking mode uses more heuristics to facilitate faster thinking and is associated with irrational decisions. People with high theory of mind intelligence are able to assess more effectively the information content of other traders' orders and so they make more successful investment predictions. As such, all three forms of intelligence have essential implications for financial decisions.

Cognitive Aging and Diminished Decision Making

Cognitive decline due to aging begins in early adulthood and slowly continues throughout life. However, financial experiences also begin early in life and increases rapidly through the middle years. This combination of declining cognitive ability and increasing financial experience results in young adults being prone to making mistakes due to a lack of experience, while seniors make mistakes because of cognitive decline. As such, peak financial decision making occurs in the age range from the forties to the early fifties and often severely impacts seniors' ability to manage investments after the age of 70. Chapter 13 illustrates how this cognitive decline affects financial decisions. Seniors are mostly aware of their cognitive decline, yet irrationally remain confident about their ability to manage their finances. Regrettably, this cognitive decline makes the elderly more susceptible to financial fraud and prone to poor investment decisions. Cognitive decline also impairs a senior's ability to manage emotions and behavioral biases in the investment decision process.

The impact of these cognitive outcomes has important ramifications for people individually. However, there are also significant societal implications given that the U.S. baby boom generation represents nearly 74 million people who are retiring at a rate of about 10,000 per day. Due to cognitive aging, they face increasing cognitive limitations to their financial decision-making abilities, yet do not realize this limitation. As this generation is the wealthiest one in history, their investment decisions and biases will have ramifications on society and the economy as a whole.

A Possible Future

Technology and big data applications are transforming how people live their lives. Genetic testing is already informing people about their inherited health risks in time to change their behavior and lower those risks. Could biological information help investors?

Consider Robert. He wears a fitness watch that tracks his heart rate and activity level throughout the day. He also has a wellness app on his phone to record what he eats, when he sleeps, and how much he weighs. His doctor also provides his health status data from his annual physical appointment and associated lab work. Robert has also been genetically tested for mapping his DNA. Given the link between biology and financial decision making, could all of Robert's data on physiological conditions such as exercise, diet, sleep, health, genetics, as well as the estimated forces on mood such as the weather forecasted, winning status of his favorite sports teams, and whether the daylight is getting shorter or longer, be combined in a Bio-Invest app that uses all of these results to tell him what behavioral biases could influence his investing choices? As such, this Bio-Invest app could inform him when he should make a financial decision, and help Robert to overcome biases induced from his DNA, current health, emotions, and mood. By overcoming his behavioral and biological biases, Robert optimizes his financial decisions and maximizes his wealth.

Unfortunately, the Bio-Invest app does not exist. However, all the biological data sources are available. It is the research gaps that are missing. This is most likely due to the interdisciplinary nature of the work. While there is much biology of investing research available on many of these particular topics, there is little on how they might integrate.

Notes

1 John R. Nofsinger. 2018. *The Psychology of Investing*, 6e. New York, Routledge.
2 Meir Statman. 2005. Normal investors, then and now. *Financial Analysts Journal* 61:2, 31–37.
3 John Gray. 1992. *Men Are from Mars, Women Are from Venus*. New York, HarperCollins.

Section I

Nature versus Nurture

2 Genetics

Twin Investment Behavior

In 1979, Jim Springer was 39 years old. He had married his first wife, Linda, and named his first-born son James Alan and his dog Toy. He and Linda later divorced, and he married Betty.[1] That is when he met Jim Lewis, the identical twin brother he never knew he had. Jim Lewis also married a Linda, named a son James Allan and a dog Toy, divorced and then married a Betty. They had both been adopted and raised 40 miles apart in Ohio. Moreover, both had worked as part-time deputy sheriffs. Clearly, there are many weird similarities between these two men who had been raised apart. Is this just coincidence, or does genetics partially drive decision making? These "Jim Twins" inspired much research on the behavior of twins who have been reared apart.

What about financial decisions? Are your investing decisions derived from the education and experiences you have had? Or are they pre-programmed through your genetic code? This "nature versus nurture" question has been examined for decades in various ways. Recently, scholars have examined the degree to which genetics, childhood experiences, and adult experiences influence financial choices. One method for exploring this topic is the use of data from twin studies. Twins share a portion of their genetic code (nature), a similar upbringing (nurture), and different adult environments (unique experiences). Through clever statistical methods and a large sample of twin investment choices, scholars can tease out estimates of the underlying sources of financial decisions.

Twin Research: Nature, Nurture, and Unique Experiences

The social sciences have been studying twin behavior for many decades to examine similarities in health, social activities, values, and religious activities, etc. These studies exploit the fact that identical twins (monozygotic twins) come from one egg and one sperm. Thus, identical twins share 100 percent of their genetic material. Alternatively, fraternal twins (dizygotic twins) come from two eggs and two sperm. Fraternal twins share on average 50 percent of their genetic material, and can be of different genders. The intuition behind twin research is that if the financial decisions of identical twins are more closely correlated than the decisions of fraternal twins, then those choices can be partially attributed to a genetic factor. When twins are raised together, they have the same shared nurturing experience. However, they will have different adult experiences. A large sample of identical and fraternal twins provides variations in nature, nurture, and unique experiences. Using econometric analysis, scholars can estimate the portion of a financial decision that can be attributed to each of these three sources.

The seminal work for twin research in economics is Paul Taubman's study that investigates the contribution of genetics to a person's earnings later in life.[2] The twin sample he used was originally assembled to study a wide variety of diseases. All of the twins in the data sample were born during the period 1917–1927, served in the U.S. armed forces during World War II, and survived that war. Given these parameters, the twins were all men. A survey was sent to the twins asking about their earned incomes, which also included questions that attempted to determine whether they were identical or fraternal twins. The study estimated that genetics explained 18–41 percent of their incomes at about the age of 50. This sample of twins has many selection biases and the participants have mostly died. Thus, most modern twin studies use a database assembled in Sweden.

The Swedish government has kept track of all twins born in Sweden since 1886. The resulting Swedish Twin Registry is the most common data source used for this type of research. The data identifies tens of thousands of identical and fraternal twins. These twins have been asked to complete telephone surveys, mail surveys, and internet surveys to discover demographic and lifestyle information about them. In order to examine financial and economic decision making, scholars in Sweden have been given permission to access and merge this twin data with other national databases such as those maintained by the military, the Premium Pension Agency, and the Swedish Tax Agency. A smaller twin database comes from the Minnesota Center for Twin and Family Research. The findings of studies using these databases, additional surveys, and lab experiments are described in this chapter.

Saving and Homeownership

Saving

Even among people of similar lifetime earnings, there is enormous dispersion in retirement wealth. Economists have been puzzled because this variation is not explained by investment asset allocation choices or socioeconomic characteristics like education. People with higher retirement wealth simply started saving earlier, which means they consumed less. Still, why do some people have a propensity for early saving while others do not? Economics characterizes the "save/consume choice" as personal discount rates. The discount rate simply characterizes how much a person values spending money now versus saving for later. People with high personal discount rates consume more and people with low discount rates save more. What is the source of this discount rate framework? Does biology partially drive this urge to save or consume?

Cronqvist and Siegel examine the fundamental origins of savings behavior.[3] They used data from the Swedish Twin Registry and matched the data with information from the Swedish Tax Agency and other data sets. Prior to 2006, Sweden had an annual wealth tax of 1.5 percent. Along with personal income, each citizen reported assets owned each year to the Tax Agency. Thus, the Tax Agency data contains everyone's portfolio holdings. By limiting their sample to twins of pre-retirement age, twins with a positive net worth, and other characteristics, the sample consisted of 14,930 twins. Ideally, savings behavior should be measured by the portion of income that is saved. Empirically, the study measured the person's change in net worth between 2003 and 2007, excluding investment gains and losses, scaled by that person's disposable income over the period. Using econometric techniques, the authors decompose the variation in savings rates into nature, nurture, and unique experiences components. The study reported that:

- Genetic differences explained about one-third of the variation in savings behavior among those studied.
- The ability of parenting (nurture) to explain the variation in savings rates was highly dependent on age. For young people, nearly half of the variation in savings rates was explained by nurture experiences. That figure declines to less than 10 percent for 45 year-olds.
- The savings rate was more dependent on a person's unique experiences in adulthood as they become steadily older.

Does genetics drive saving/consumption directly, or does it influence related behaviors that affect savings? The study showed that genetics was correlated with income growth, smoking, and obesity. Thus, genetics could impact time preferences and self-control, which then impact saving and consumption.

Homeownership

In addition to saving and investing, another source of retirement wealth is homeownership. Cronqvist and colleagues investigated whether homeownership may be partially driven by biology.[4] They used the Swedish Twin Registry and merged it with various government data for homeownership and location, resulting in a sample of 26,876 twins. They studied the influence of nature, nurture, and unique experiences on the decision to own a home and the value of that real estate.

The study reported that 21 percent of the variation in the decision to own a home can be attributed to genetics. One-quarter of the variation in home values can also be explained by biology. The nurturing shared experiences among the twins did not influence the homeownership and value decisions.

Thus, the main source driving these decisions was people's unique experiences. Further analyses showed that the shared nurture environment does influence these real estate choices for younger people (age < 35), but that influence dissipated to zero for middle-aged and older people. Unique experiences exerted more influence on these decisions as people aged. Biology appeared to have a significant influence over the entire life cycle.

Risk Attitudes

Risk aversion plays a very important role in the theory and application of financial decisions. The traditional view of risk aversion is an important component of evaluating investment alternatives in risk-return models, such as Markowitz's mean-variance portfolio theory. In traditional economics, risk aversion is determined through the utility function within the expected utility theory. In this framework, people only take risks when they offer a sufficient reward, known as a risk premium. For example, would you prefer a certain payoff of $5,000 or a gamble with a 70 percent chance of getting $10,000 and a 30 percent chance of getting nothing? Note that the gamble has an expected return of $7,000 (=0.7 × $10,000 + 0.3 × $0), which is $2,000 higher than the $5,000 in the certain choice. This $2,000 represents the risk premium of the gamble. The size of the risk premium needed to induce you to take the risk is an indication of your risk aversion. People with a higher risk aversion need a higher premium to take risks. Given a choice between several alternatives with different levels of risk and reward, people choose the one that offers the highest utility. Alternatively, the competing explanation is the behavioral economic theory of economic risk called prospect theory, which shows that people are subject to cognitive biases that cause them to be risk averse when it comes to gains but risk seeking when it comes to losses. Determining risk preferences is important for both economic theories. While many studies have examined the consequences of risk preferences in both theories, little work has been done on the source of risk preferences. Might genetic factors influence risk preferences?

Zyphur and colleagues examined heredity in risk preferences using a sample from the Minnesota Twin Registry, which includes male twins raised together and born between 1961 and 1964.[5] The twins were asked to complete a survey that included questions designed to determine their risk preferences. The final sample included 400 people, 111 identical pairs and 89 fraternal twin pairs. Their risk preferences were determined by combining answers from three questions into one risk measure. The three questions were:

Question 1: Which of these three opportunities would you prefer?

A   Win a $2,000 prize

B   Take a gamble in which there is 50 percent chance to win $5,000 and a 50 percent to win nothing

    C Take a gamble in which there is a 20 percent chance to win $15,000 and an 80 percent chance to win nothing

Question 2: You are making an investment in your retirement plan. Retirement is 15 years away. Which do you prefer?

    A A money-market fund or guaranteed investment contract, giving up the possibility of major gains, but virtually assuring the safety of your principal

    B A 50–50 mix of bond funds and stock funds, in the hope of getting some growth, but also giving yourself some protection in the form of steady income

    C Aggressive growth mutual funds whose value will probably fluctuate significantly during the years, but have the potential for impressive gains in the long term

Question 3: Your employer—a private entity—is selling stock to employees. Management has plans to take the company public in roughly three years. Until the company is taken public, there is no possibility of selling shares or receiving dividends, but if the company goes public, increases in stock valuation could be quite large. How much would you like to invest in the company stock?

    A None
    B Two months' salary
    C Four months' salary

The study concluded that genetics explained 63 percent of the variability of risk preferences among people. The balance, 37 percent, was attributed to unique experiences with no contribution from shared nurture experiences.

    Cesarini and colleagues invited Swedish twins from the Swedish Twin Registry to participate in a survey to examine risk preferences.[6] A total of 460 twin pairs (141 identical pairs and 319 fraternal pairs) came to the experimental setting. Risk preferences were measured using monetary incentives in two ways, labelled risk aversion and risk investment. For the first, risk aversion, the twins were presented with six questions that each had the same gamble but different certain outcomes. The gamble was a 50 percent chance to win 100 Swedish Krona (SEK) and a 50 percent chance to win nothing. The six questions were:

    1 Do you prefer a certain SEK 20, or a 50–50 gamble for SEK 100?
    2 Do you prefer a certain SEK 30, or a 50–50 gamble for SEK 100?
    3 Do you prefer a certain SEK 40, or a 50–50 gamble for SEK 100?
    4 Do you prefer a certain SEK 50, or a 50–50 gamble for SEK 100?
    5 Do you prefer a certain SEK 60, or a 50–50 gamble for SEK 100?
    6 Do you prefer a certain SEK 80, or a 50–50 gamble for SEK 100?

Note that if a person is willing to accept a certain SEK 40 over the gamble, then they would be willing to accept a certain SEK 50, SEK 60, and SEK 80 too. Thus, the questions provide the crossover point for which a person changes from taking the gamble to the certain payoff. That switchover point is an indication of the level of risk aversion. The lower the minimum certain payoff a person is willing to accept in order to avoid the gamble, the higher their risk aversion.

Risk investment was assessed with the scenario that the twins had won SEK 1 million in a lottery. They could invest some of this money in a risky asset with a 50 percent chance of doubling it and a 50 percent chance of losing half the investment. The subjects then selected the amount they wanted to invest among the choices: SEK 0, SEK 200,000, SEK 400,000, SEK 600,000, SEK 800,000, or SEK 1 million. The higher the amount selected, the greater the level of risk investment.

The study showed that the correlation between identical twins' risk aversion was 0.222, compared to 0.025 for fraternal twins. For risk investment, the correlations were 0.264 and 0.096, respectively. The main difference between identical and fraternal twins is that identical twins share a higher proportion of their genetic code. Since the correlation for identical twins was significantly higher than for fraternal twins, the genetic code is likely to play an important role in their risk preferences. In decomposing the variation in risk preferences between nature, nurture, and unique experiences, the authors reported that genetics explained 14 percent of risk aversion, while shared experiences and unique experiences explained 7 percent and 80 percent, respectively. Risk investment was attributed to genetics (19 percent), shared experience (10 percent), and unique experiences (71 percent). Both estimates for the genetic factor (14 percent and 19 percent) are much lower than the 63 percent reported by Zyphur and colleagues. Nevertheless, both studies reported that risk preferences are heritable traits.

Investing

Stock Market Participation

Barnea and colleagues combined Swedish Tax Agency portfolio data with a sample of over 37,000 twins and an equal sample of non-twins.[7] Presumably, each asset held in the portfolio was once a conscious decision to purchase it. They examined the investment portfolios in three ways. First, had the person invested in the stock market? Second, what portion of their investment portfolio was invested in the stock market? Finally, how risky was the portfolio as measured by volatility? Then they compared these investment portfolio characteristics between the various categories of twins and non-twins using correlation analysis. One set of correlations is shown in Figure 2.1 for whether the person has invested in the stock market.

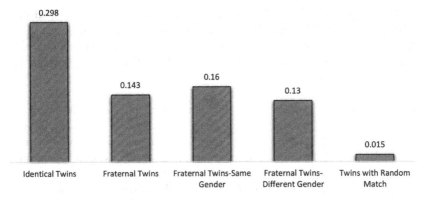

Figure 2.1 Decision Correlation between Groups of Twins

Remember that the early nurture experiences of identical and fraternal twins are commonly shared. If decisions are driven by this nurturing, then it does not matter if the twins are identical or fraternal—the correlation between the twins should be the same. As shown in Figure 2.1, the correlation between identical twins is twice that between fraternal twins for stock market participation. Since the important difference between identical and fraternal twin behavior is the portion of their genetics that they share, this result suggests that nature drives an important portion of this investment decision. The authors found similar results for the other two portfolio questions. That is, the correlation for identical twins in the portfolio proportion in equity (risk volatility) was 0.307 (0.394), which was twice the correlation of 0.150 (0.181) for fraternal twins.

Note from Figure 2.1 that there is a difference in correlation between same sex fraternal twins and opposite sex twins. The correlation for the same sex twins is higher, which means they invested more similarly than opposite sex twins. This is indicative of the gender differences in investing, which are explored in Chapter 4. Finally, note that as might be expected the correlation for stock market participation between a twin and an age-matched random person is essentially zero.

Using econometric techniques, these scholars estimated the proportion of nature, nurture, and unique experiences that drove investment decisions. The statistical models controlled for individual characteristics, like age, income, gender, wealth, education, etc. The study found that genetics explained nearly 30 percent of the decision differences seen between people. This is a particularly large number considering the personal characteristics that could drive investment decisions. For example, there is an old investment adage that an investor's equity allocation should be 100 minus his or her age. Thus, stock market participation should be related to a person's age. Also, people of higher wealth might invest more in the stock market. However, the authors reported that the proportion explained by genetics was larger than characteristics like age, gender, education, and wealth combined! Surprisingly, this study reported that shared environment, i.e., nurture, explains very little of these investment decisions. They concluded that a person's genes and adult experiences mattered a great deal more than how he or she was raised. Additional analyses reported these subsample findings:

- Age was an important factor. For younger people (age < 30), genetics explained 44.5 percent of the stock market participation decisions. Genetics explained 19.2 percent of the decisions for middle-aged people. Genetics explained even less for seniors. This makes sense because over time people gain more experiences that drive their investment decisions.
- Twins who had more contact with each other during adulthood showed more similar investment decisions. Indeed, Calvent and Sodini conducted a similar twin study and concluded that some of the proportion of a decision attributed to genetics may come from communication between the twins.[8]
- Gender matters. Genetics explained 29.1 percent of the stock market participation decision for men and only 22.1 percent for women.
- Twins reared apart showed a higher proportion of the stock market participation decision attributed to genetics (at 38.5 percent) than twins reared together (at 26.2 percent). This was likely due to the fact that nurture for twins reared apart is not shared, and thus cannot explain investment decisions.

Pension Choices

Prior to 2000 the retirement pension allocation investments for Swedish citizens was determined by the Swedish Premium Pension Agency. In 2000, the government divested this responsibility to each person. The new system provided nearly 500 investment funds from which each person needed to select for inclusion in their own retirement portfolio. People were allowed to select up to five funds. Each fund's level of risk was color-coded for easy interpretation. For the first time, Swedish citizens had to allocate their own retirement portfolio. Was genetics a factor in their decisions? Cesarini and colleagues studied this question by investigating this mandatory 'handover' of investment choices in the national pension system using the decisions made by twins.[9] There are several advantages to examining this initial allocation decision. Studies using existing portfolios may be biased because portfolios can be distorted over time when one asset class outperforms the others. If an investor does not actively rebalance the portfolio periodically, the original investment choices may be obscured by how varying asset returns distort the portfolio's asset allocation over time. For example, if an investor allocates 50 percent of their assets to stocks and stocks outperform bonds and cash, then five years later 70 percent of the portfolio might comprise stocks. The investor's decision was to allocate 50 percent to stocks, but the portfolio shows a 70 percent allocation to stocks. So, this study examined the actual allocation decision.

For people who did not want to make this investment decision, there was a default option. About 32 percent used the default option. The other 68 percent made an active choice. This nationwide policy change allowed for a unique opportunity to examine how decisions may be attributed to nature, nurture, and unique experiences. Using econometric techniques, the authors reported many interesting findings:

- Genetics explained 28 percent of the level of risk taken.
- Shared rearing environment (nurture) explained very little.
- Some of the funds offered advertised themselves as socially responsible funds. Genetics explained a very large portion (60 percent) of the decision to select these funds.
- Analysis of the fund choices showed that some had performed very well lately. One psychological bias is to extrapolate past returns to future expectations. Specifically, they identified the funds whose returns were in the top 10 percent of their category during the last three years. People often exhibited this extrapolation bias by selecting these funds. Genetics explained 30 percent of the decision to chase these high-return funds.

Investing Style: Value versus Growth

Much of the investment industry is described by investment style. One popular dimension is the value versus growth continuum. Value investors look for stocks that appear to trade at a lower price than their intrinsic value, whereas growth investors look for firms with high earnings or sales growth. Investors are often categorized as being value investors or growth investors. The mutual fund industry advertises many of its funds as being either value or growth oriented, while Morningstar evaluates those characteristics in the funds. Academics noted the differences in returns between value and growth stocks and generated the value premium debate to determine whether the premium is an anomaly or compensation for risk. But who prefers value-oriented investing and who prefers growth investing? Does this preference stem from genetics or experiences? For example, Benjamin Graham is commonly referred to as the father of value investing, while T. Rowe Price, Jr. is often referred to as the father of growth investing. We will never know if biological differences drove them to value and growth. However, they did have very different upbringings. Graham grew up poor. His father died when he was young and his mother lost the family savings in the stock market panic of 1907. Price grew up in a wealthy family with a father who worked as a surgeon for a railroad growth company.

Cronqvist and colleagues determined to find the source of investing styles using the Swedish twin data.[10] In order to assess a person's value versus growth tilt, the study first characterized each stock's degree of value or growth through its price/earnings ratio and its price/book ratio. Mutual funds were categorized by Morningstar's Value-Growth score. Then, each person's value-growth tilt was assessed by examining the portfolio holdings and combining the assets' value-growth measures into one measure. Once the value-growth tilt of each person was computed, the researchers determined the source of this tendency. In short, they argued that differences in value-growth style stems from both biological predispositions and environmental factors.

Specifically, over 25 percent of the preference for value versus growth stocks was explained by genetic differences. What personal demographics or experiences drive style preference? The study reported that:

- People with more education and higher incomes had more growth-oriented portfolios, like T. Rowe Price, Jr.
- People who grew up in a lower socioeconomic status had a stronger value orientation, like Benjamin Graham.
- Younger investors tilted toward growth.
- People who grew up during the Great Depression had more value-oriented portfolios.
- People who entered the labor market during an economic recession were more value oriented.

Behavioral Biases

There are many investment biases. Some are more easily examined through the trades and portfolio holdings of investors, while others can be assessed through survey questions or lab experiments. Two studies investigate the role of nature and nurture in the exhibition of behavioral biases. Both studies use the Swedish Twin Registry, but one implements a survey of the twins while the other examines the portfolios revealed by the Tax Agency.

In the first study, Cesarini and colleagues offered a survey to the twins.[11] While surveys have advantages and disadvantages as a research tool, the advantages are that hypotheses can be more directly tested, and the environment can be better controlled. The survey was completed by 3,512 sets of twins. Seven psychological biases were examined: representativeness bias, loss aversion, illusion of control, sunk cost bias, status quo bias, procrastination, and time impatience. Using the same econometric techniques as those in earlier studies, the researchers estimated that biology did play a role in revealed psychological biases. Figure 2.2 shows that the portion of a bias explained by genetics varies from one-fifth to one-third depending on the specific bias. The largest role for genetics was estimated for the sunk cost bias, which was measured using a question about going to a show after the ticket was lost. The next greatest genetic influence was found for the representativeness bias, which was measured using three questions about the likelihood of people belonging to different groups. Genetics was found to explain almost one-quarter succumbing to three biases: loss aversion, illusion of control, and status quo bias. Loss

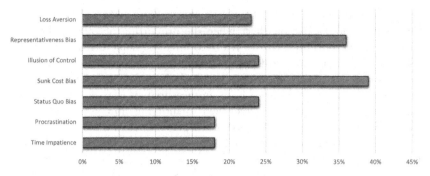

Figure 2.2 Portion of the Psychological Bias Explained by Genetics

aversion was measured through three lottery questions, while illusion of control was evaluated through questions about the discount acceptable to choose your own lottery numbers versus getting random numbers. Status quo bias was determined by a question about switching to a new and cheaper service provider. Genetics was estimated to have the least impact (a portion of a little less than one-fifth) on procrastination and time impatience. Procrastination was appraised through how often bills were paid late. A question about the discount acceptable to receive money sooner gauged time impatience. The researchers also reported that most of the explanation for exhibiting behavioral biases appears to derive from people's unique experiences and that very little came from the nurture environment.

In the second study, Cronqvist and Siegel merged the twin data with the portfolio holdings from the Tax Agency data.[12] The portfolio holdings are recorded annually. By comparing a person's portfolio from one year to the next, the scholars can determine what was purchased or sold. However, the timing of those trades and the prices for which they were executed were not known. Through examination of the portfolio itself, the researchers measured the degree of three investment biases: diversification, home bias, and skewness preference. Under-diversification is often attributed to overconfidence. The proportion of the portfolio invested in Swedish securities is a measure of the home bias—a ramification of familiarity bias. Skewness preference is a bias toward holding stocks with gambling characteristics and was computed as the proportion of the portfolio invested in "lottery" type securities. Stocks with lottery characteristics have a low price and a high risk (volatility).

Three psychological biases were computed using the changes in the portfolio from one year to the next: turnover, the disposition effect, and performance chasing. Turnover is a measure of the degree of trading. Excessive trading is a ramification of overconfidence. Turnover was calculated as the number of securities sold in relation to the total number of securities held. The disposition effect is related to loss aversion and is the propensity to sell winning positions and to hold losing positions. From the portfolio holdings the authors estimated the gain in the securities held and sold for winners and the realized and unrealized losses in the securities sold and held. Investors exhibit the disposition effect when the proportion of gains realized is larger than the portion of losses realized. Finally, people often extrapolate recent stock performance to future expectations. Thus, this form of representativeness bias (also known as extrapolation bias) urges investors to chase high past performance. This was measured by classifying all assets into return deciles each year and then quantifying how well the stocks purchased by an investor performed in the previous year.

After controlling for demographic characteristics like age, education, wealth, etc., the underlying sources that drove these investor behaviors were decomposed into proportions from genetics, shared upbringing, and unique experiences. The study reported that the shared upbringing (nurture) contributed very little to these revealed biases. Genetics (nature) and unique experiences drove these investment choices:

- Diversification: genetics 45.3 percent, unique experiences 51.6 percent, shared upbringing 3.1 percent.
- Home bias: genetics 45.2 percent, unique experiences 54.8 percent.
- Turnover: genetics 25.1 percent, unique experiences 74.9 percent.
- Disposition effect: genetics 27.2 percent, unique experiences 72.8 percent.
- Performance chasing: genetics 31.1 percent, unique experiences 59.4 percent, shared upbringing 9.5 percent.
- Skewness preference: genetics 27.5 percent, unique experiences 72.5 percent.

Note that several psychological biases were examined in both studies. For example, the loss aversion in the first study examined a similar idea to the disposition effect used in the second study. While the research methods were very different (survey versus holdings analysis), the estimates for the degree to which genetics influenced this loss-averse behavior was similar at 23 percent versus 27 percent. The genetic influence for representativeness bias in the first study matched the performance chasing in the second study at 31 percent. Overall, the two studies showed that the dominant influence was the unique experiences followed by shared biology.

Well-being: Giving and Happiness

Giving

To examine altruism, Cesarini and colleagues invited Swedish twins to participate in a lab experiment.[13] This study also examined risk preferences and was mentioned earlier. The experiment examined the twins' propensity toward giving. A total of 460 twin pairs (141 identical pairs and 319 fraternal pairs) came to the experimental setting. To measure giving propensity, the twins participated in a modified form of the dictator game. In the dictator game, a participant decides how to split a sum of money between himself and another person. The modification was that the participant splits the money between himself and a charity. In this experiment, each twin divided $15 into a portion they could keep, and a portion to be donated to a charity for the homeless. The study reported that genetics explained 28 percent of the variation in giving, while nurturing and unique experiences explained 11 percent and 61 percent, respectively.

Happiness

Economists try to explain the determinants of subjective well-being or life satisfaction. Economic factors can include personal variables (like income, socioeconomic status, marriage, education, and religiosity) and society variables (like democracy, inflation, and the unemployment rate). However, these factors account for only a small variation in people's happiness. Changes in these factors have only a short-term impact on the level of happiness. A person's happiness level appears to revert to a "baseline" level. This baseline is high for some people and low for others. The baseline is partly formed by the person's personality and genetic predisposition. In short, people are biologically predisposed to a general level of happiness or sadness. That level of baseline happiness is significantly heritable.[14]

Using a twin study, De Neve and colleagues estimated the genetic contribution to life satisfaction.[15] The scholars utilize the Add Health project, which was originally created to study the health-related behaviors of adolescents. It has been expanded and used in many fields. The first wave of survey questions of high school aged children and younger occurred in 1992–1994 and included tens of thousands of students. Three additional waves of surveys were conducted through 2008 that followed up with approximately 15,000 students. This sample included 434 identical twins and 664 fraternal twins. The third wave of surveys included a question on happiness. Specifically, it asked, "How satisfied are you with your life as a whole?" Possible answers were very dissatisfied, dissatisfied, neither satisfied nor dissatisfied, satisfied, and very satisfied. The life satisfaction answer for identical twins had a correlation of 0.334, which was significantly higher than the correlation of 0.132 for fraternal twins. The difference shows that identical twins were more similar in their level of well-being than fraternal twins, which indicates that biology might play a role in happiness. In decomposing the variation of the well-being responses, the study found that 33 percent could be attributed to genetics, 67 percent was attributed to people's unique experiences, and shared nurture did not contribute. Additional analysis used a similar question asked in the first and fourth waves, "How often was the following true during the past seven days? You felt happy." The estimated genetic contribution to happiness in waves one, three, and five were 22 percent, 33 percent, and 54 percent, respectively. This indicates that genetics became more important in setting baseline happiness as the person grew older.

Summary

Twin siblings are of the same age and have been raised together in the same family. Yet genetically identical twins still exhibit much greater similarity in their preferences for saving, buying houses, risky investing, behavioral biases, giving, and happiness than do fraternal twins. The environment in which they were raised appears to drive a surprisingly small portion of these decisions. Genetics seem to explain a significant amount of the variation in financial decisions. Depending on the decision, biology is found to explain one-fifth to one-half of the variation. In that regard, nature drives our decisions more than nurture. The unique experiences a person has explains the largest variation in decision making. The shared experiences of being raised together has some impact in the decisions for younger adults, but that influence declines to zero over time. In its place, the unique experiences of adulthood explain more decisions over time. The genetic coding for decisions appears to influence a person throughout their life.

Questions

1 What are the fundamental questions and issues in the nature versus nature debate?
2 Why do genetics studies compare identical twins' behavior to fraternal twins' behavior?
3 What can we conclude from twin studies about risk taking?
4 What does twin research say about the source of irrational behaviour such as investment biases?

Notes

1 Nancy Segal. 2012. *Born Together—Reared Apart: The Landmark Minnesota Twin Study*. Boston, MA, Harvard University Press.
2 Paul Taubman. 1976. The determinants of earnings: Genetics, family, and other environments: A study of white male twins. *American Economic Review* 66:5, 858–870.
3 Henrik Cronqvist and Stephan Siegel. 2015. The origins of savings behavior. *Journal of Political Economy* 123:1, 123–169.
4 Henrik Cronqvist, Florian Münkel, and Stephan Siegel. 2014. Genetics, home-ownership, and home location choice. *Journal of Real Estate Finance and Economics* 48:1, 79–111.
5 Michael J. Zyphur, Jayanth Narayanan, Richard D. Arvey, and Gordon J. Alexander. 2009. The genetics of economic risk preferences. *Journal of Behavioral Decision Making* 22:4, 367–377.
6 David Cesarini, Christopher T. Dawes, Magnus Johannesson, Paul Lichtenstein, and Björn Wallace. 2011. Genetic variation in preferences for giving and risk taking. *Quarterly Journal of Economics* 124:2, 809–842, and associated Online Appendix.
7 Amir Barnea, Henrik Cronqvist, and Stephan Siegel. 2010. Nature or nurture: What determines investor behavior? *Journal of Financial Economics* 98:3, 583–604.

8 Laurent E. Calvent and Paolo Sodini. 2014. Twin picks: Disentangling the determinants of risk-taking in household portfolios. *Journal of Finance* 69:2, 867–906.

9 David Cesarini, Magnus Johannesson, Paul Lichtenstein, Örjan Sandewall, and Björn Wallace. 2010. Genetic variation in financial decision-making. *Journal of Finance* 65:5, 1725–1754.

10 Henrik Cronqvist, Stephan Siegel, and Frank Yu. 2015. Value versus growth investing: Why do different investors have different styles? *Journal of Financial Economics* 117:2, 333–349.

11 David Cesarini, Magnus Johannesson, Patrik K. E. Magnusson, and Björn Wallace. 2012. The behavioral genetics of behavioral anomalies. *Management Science* 58:1, 21–34.

12 Henrik Cronqvist and Stephan Siegel. 2014. The genetics of investment biases. *Journal of Financial Economics* 113:2, 215–234.

13 David Cesarini, Christopher T. Dawes, Magnus Johannesson, Paul Lichtenstein, and Björn Wallace. 2011. Genetic variation in preferences for giving and risk taking. *Quarterly Journal of Economics* 124:2, 809–842.

14 David Lykken and Auke Tellegen. 1996. Happiness is a stochastic phenomenon. *Psychological Science* 7:3, 186–189.

15 Jan-Emmanuel De Neve, Nicholas A. Christakis, James H. Fowler, and Bruno S. Frey. 2012. Genes, Economics, and Happiness. *Journal of Neuroscience, Psychology, and Economics* 5:4, 193–211.

3 More Genetics

Investments of Adoptees and the Human Genome

How do behaviors pass from one generation to the next? Is it due to parents contributing their genome to their children, or is it due to the rearing environment they provide? The previous chapter reviewed studies that exploited the differences in the known amount of shared DNA between identical and fraternal twins to assess the contribution of nature, nurture, and unique experiences in financial decision making. However, these studies have received some criticism that too much of the variation between identical and fraternal twins is attributed to genetics. Critics argue that identical twins communicate with each other throughout their lives more than fraternal twins do and that some of their communication could be about investing. Therefore, it is valuable to look for other ways to examine the nature versus nature question.

This chapter examines two other approaches. The first is to examine adoptee behavior. Adoptees share none of their DNA with their adopted parents and much DNA with their biological parents. Yet their rearing experiences are from the adopted parents. Thus, much can be learned about nature and nurture through examining adoptee economic preferences and comparing them to those of their biological and adoptive parents, and adoptive siblings.

The second approach is to exploit genetic markers in DNA directly. The completion of the Human Genome Project in the early 2000s was followed by significant cost reductions in mapping a person's genome structure. As a result, commercial DNA genetic testing companies (like AncestryDNA, 23andMe, Genos, and Veritas Genetics) offer to genotype people in order to learn about their biological history and health predispositions. Also, major funding bodies have incorporated genotyping into national social surveys in several countries. It is now possible for scholars to directly examine the genetic variants that may predict differences in economic preferences and outcomes.

Adoption

The twin studies distinguished between nature, nurture, and unique experiences by decomposing the variation in investor behavior using the known variation in shared genes and by making strong assumptions about the shared rearing experiences of the twins. Some of the adoption literature also uses this method. However, some of the adoption studies use a different approach, such as regression analysis, to estimate parameters for the relative importance of the similarities between adoptee behavior and the behavior of biological and adoptive parents. While this approach relaxes the strong assumptions about similarities in siblings' shared experiences, it does add the assumption that adoptive parents are randomly assigned for adoptees. This condition may not be true as adoptive parents tend to be a little better educated and wealthier than non-adoptive parents and a lot wealthier than the adoptive child's biological parents. Finally, the adoptive regression methods do not separate childhood rearing experiences from unique adult experiences. Thus, this literature comments only on the relation between contributions from genetics and environment. That is, adoptive research comments on the importance of nature relative to the combined nurture and unique experiences. These are known as post-birth effects. Regardless of the method used, adoptee research has determined that genetics plays a significant role in a person's entrepreneurship,[1] voting behavior,[2] longevity,[3] criminal convictions,[4] and educational attainment.[5]

Introduction to Adoption Studies: Education and Income

Following the Korean War, many orphans in South Korea were adopted by citizens in the United States and other countries around the world. In the United States, the Holt International Children's Services facilitated the adoption of tens of thousands of children. The process included considerable vetting of the would-be American parents, but the children were adopted on a first-come, first-served basis, thus making the process quite random. Sacerdote surveyed the parents of adopted Korean Americans about the adoptees' lives and the lives of the parents' non-adopted children.[6] The survey sought information from the parents, adoptees, and non-adoptee children about their income, education, health, etc. The author exploited the fact that adoptees did not share any genetic markers with the family, while the non-adoptee children did. The study used the variance decomposition methods used by the twin studies to allocate the influence of nature, nurture, and unique experiences to various attributes and decisions of the children. While the study was not investment related, the results allow for the assessment of whether adoptee research produces results that make sense. For example, consider the source of the child's "height" variable shown in Figure 3.1. One would expect a person's height to be strongly dependent on the height of the biological parents, not the height of the adoptive parents or the nurture experience. The results were consistent with this conjecture; nearly 86 percent of a person's height was attributable to genetics, as the remaining proportion could be explained by other factors such as diet and environment.

Now consider a person's weight. It seems reasonable that genetics plays a role but that learned eating and exercise habits acquired during the rearing years, as well as unique experiences in adulthood, should also be influential.

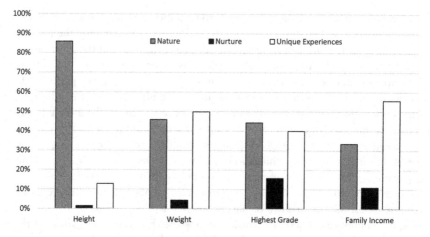

Figure 3.1 Variance Decomposition Attributable to Nature, Nurture, and Unique
Experiences

The results show that nearly 46 percent of a person's weight was attributable
to genetics. About 4 percent was attributable to nurture, and 50 percent
depended on unique experiences. Again, these results seem reasonable. The
next outcome to be explained is education. What drives the highest grades
attained by an adoptee? Figure 3.1 shows that 44 percent of the education
outcome was attributable to genetics. Nurture and unique experiences
explained 16 percent and 40 percent, respectively, suggesting that genetic
intelligence plays a role, but so too does the environment in which the
adoptee grew up (i.e., a push toward higher education attainment). Finally,
Figure 3.1 shows the results for family income. Genetics was attributed to
one-third of the variation in income. Nurture and unique experiences
explained 11 percent and 56 percent, respectively possibly due to the difference
in educational attainment and growing up in a wealthier environment. Overall, the height and weight results provided reasonable estimates that provide
some confidence in the estimates for education and family income. Genetics
appears to be a significant driver of educational and job income decisions.

Additionally, another study examined education and income but used a
different data source and employed the regression method rather than the
variance decomposition method for the analysis. In addition to the twin
database identified in Chapter 2, Sweden also keeps records about adopted
children and their biological and adoptive parents. Björklund and colleagues
used this data and focused on adoptions between 1962 and 1966.[7] They
examined the adoptees' education and income levels in 1999 when they were
aged between 33 and 37 years. This data was compared to the biological and
adoptive parents' education levels and incomes. The researchers' final
sample included about 2,000 adoptees for whom both biological parents and
adoptee parents were known and matched income and education level data.

Also included in the analysis were the adoptive siblings, referred to as the adoptive parents' own-birth children.

The study reported that for own-birth children, the education of both the mother and father was equally important in impacting the education level of the child. The income of the father was also an important determinant for the child's income. Note that the regression methodology for adoptees only divides attributes into pre-birth and post-birth factors. The pre-birth factors are considered genetic, while the post-birth factors are the combination of nurture and unique experiences. For the adoptee child's education, both the biological and adoptive fathers' years of schooling were equally important. Interestingly, while both the biological and adoptive mothers' education levels were important, the biological birth mother's education was more important than that of the adoptive mother. In regard to the adoptee's income, both the biological and adoptive fathers' income was essential, but the adoptive father's income was more important than the income of the biological father. The paper specifically concluded that:

- The genetics of the biological parents matter.
- Adoptive parents' education matters for nurture.
- The mother's influence on education is stronger pre-birth (i.e., genetics).
- The father's influence on education is equally important pre-birth and post-birth.
- The adoptive father's income (post-birth effects) is important for adoptee income.

Finally, in a follow-up study, Björklund and colleagues utilize the vast amount of Swedish data to examine the importance of nature and nurture within six family types.[8] These six types provide different combinations of biological and adoptive parents present to rear the child. The types are (1) child reared by both biological parents; (2) child reared by biological mother with no father present; (3) child reared by biological mother with stepfather; (4) child reared by biological father with no mother present; (5) child reared by biological father with stepmother; and (6) child reared by adoptive mother and father. Note that in some samples the biological father was raising his child (types 1, 4, and 5) while the biological mother was raising her child in other samples (types 1, 2, and 3). In other family structures, step- or adoptive parents were involved. This allowed for much variation in the data with a sample size ranging from 181 (type 5) to 26,596 (type 1).

The results of this study are remarkable. In the case of parents rearing their own-birth children (type 1), the education of both parents was highly and equally important. Note that for the regression methods used, this captures both the pre-birth (genetics) and post-birth (nurture and unique experiences) contributions. It is not a surprise that the mother's education was important when she raises her own-birth child by herself (type 2) or with a stepfather (type 3). In both cases, the biological father's education was also just as important as the mother's education. However, the stepfather's education was less important. A similar conclusion was reached from fathers raising their own-birth child alone (type 4) or with a stepmother (type 5). That is, the father's education was important, but so was the biological mother's education, while the stepmother's education level was less important. Finally, for the adoptee sample (type 6), the biological father's education level was equally important as the adoptive father's education while the biological mother's education was more important than the adoptive mother's education level. In summary, the biological parents' education levels were always important—sometimes more important than step- or adoptive parents' levels of education.

These three studies used two different data sources, databases containing details of Korean American and Swedish adoptions, and employed three different methodologies. All three concluded that genetics was important for influencing decisions about education and outcomes like income, thus suggesting that genetics should also be important in other aspects of decision making such as finance.

Investing Decisions

Black and colleagues were able to identify 3,185 adults who had been adopted and to match them with their adoptive and biological parents using the Swedish adoption database.[9] They examined how similar the adoptees' investment choices were to both sets of parents. This study of adoptees provided a situation in which nurture occurs in the absence of common genetics. In addition, they looked at how similar investment choices were between over two million non-adoptees and their parents. Like the twin research, these scholars used the Swedish Tax Agency data on wealth to determine whether a person participated in the stock market, the portion of the portfolio invested in equity, and the risk (measured as volatility) of the portfolio. The investment decision similarities of over two million non-adoptees with their parents provides a baseline for how similar their investment decisions were. The regression analysis methods demonstrate the relative importance of genetics (nature) and environment (nurture plus unique experiences).

There may be some concern that parents treat an adopted child differently from their own-birth child, at least in such a way that might impact future investment decisions. However, the study reported that the intergenerational transmission of investment decisions to own-birth children was approximately of the same magnitude of the sum of the biological and adoptive effects together for adopted children. Thus, the influence of both genetics and nurture on investment tendencies may not be impacted by the adoption itself.

The overall findings were that genetics and environment are important in determining whether a person participated in the stock market and their level of risk aversion. Biology did not appear to influence the portion of a person's portfolio invested in equities. In general, environment (nurture and unique experiences) had twice the effect of the biological influence for stock market participation. The environment had a four-times larger influence than genetics for the level of risk in the portfolio. Another interesting finding was that analysis by gender showed that women were more influenced by their adoptive mothers' investment decisions and men by their fathers' decisions. These results suggest that the environment that children grow up in may play a more important role than their biology.

Another adoptee database comes from Norway. This data includes South Korean infants adopted by Norwegian families. Fagereng and colleagues used this data of 2,265 Korean children adopted between 1965 and 1986 to examine the intergenerational transmission of wealth and also the attitude toward risk aversion in the absence of genetic similarities.[10] As is the case in Sweden, Norwegian scholars have access to records about peoples' wealth, income, and investment assets. Specifically, the researchers examined the decision to participate in the stock market and the portion of a person's portfolio invested in equities. Researchers examined the wealth accumulation and these two measures of financial risk taking of the adoptees and their parents. They concluded that the rearing environment had a great influence. When an adoptee was raised by parents who took more financial risk, so did the adoptee.

However, this study did not have data on the adoptee's biological parents' investment behavior. So, in order to compare to the twin studies, they examined the behavior of adoptees and non-adopted siblings. They exploited the fact that the Korean Norwegian adoptees shared no biological markers with their adoptive parents, while the non-adoptive siblings did. Using the same variance decomposition methods used in the twin studies, but relaxing the assumption of a random adoption process, these authors estimated the nature, nurture, and unique experience contributions to net wealth, financial wealth, portfolio equity allocation, and educational attainment. The estimated driving forces of these attributes are shown in Table 3.1.

The authors found that nature and nurture both have a strong influence on net and financial wealth. Indeed, the genetic component attributed over 50 percent in both cases. Unique experiences had a small influence on wealth. However, for the portfolio allocation to equities, genetics did not

Table 3.1 Sources of Korean Norwegian Adoptees' Decisions

	Nature	*Nurture*	*Unique Experiences*
Net Wealth	57.6%	36.5%	5.8%
Financial Wealth	52.3%	24.6%	23.1%
Risky Portion of Portfolio	−5.5%	14.1%	91.4%
Education	49.2%	5.9%	45.1%

play a role. At an estimated 91.4 percent, unique experiences drove the portfolio allocation decision. This makes sense because an initial portfolio allocation to stocks is strongly affected by how well the stock market performed during the preceding year prior to the initial allocation decision—a representativeness bias. Since adoptees, siblings, and parents make this initial decision at different times, this bias would influence them to make different allocations. These unique experiences could be much more important than nature and nurture factors. Finally, nearly half of the level of education attained can be attributed to genetics with very little attributed to nurture. Overall, relaxing the strong assumptions of the twin models caused much of the influence attributed to unique experiences to shift to nature and nurture.

Genome

The Human Genome Project successfully mapped human DNA. Specifically, it determined the sequence of nucleotide base pairs that make up human DNA and identified and mapped the location and functionality for all the genes of the human genome. It seems that scholars should be able to genotype people and then compare their genetic makeup to their investment characteristics. However, as noted below, there are some problems to overcome.

The costs of genotyping a person continue to fall, so large-scale studies to associate specific genes with financial behavior continue to evolve. However, genotyping cost is just one obstacle to overcome. A person's genome consists of 23 pairs of unique chromosomes, one copy of each unique pair from each parent. Each chromosome is segmented into hundreds and thousands of functional regions called genes. Each gene contains potentially millions of nucleotide pairs. It is the string of these nucleotide pairs that is illustrated by DNA strands. Thus, one problem is that the human genome has approximately three billion nucleotide pairs. Determining which combinations of them influence financial decision making would be improbable with normal numbers of people in the sample. Luckily, all humans share 99.6 percent of their genetic variation. Therefore, scholars study the genetic differences between people and various kinds of decision making. Still, the differences entail millions of genetic markers. The nucleotide pairs in which people may differ are called single nucleotide polymorphisms, or SNPs. Scientists try to link these "snips" to physical characteristics and behavior. Since the number of SNPs far exceeds the number of people in any sample, the initial tests had very low statistical power to detect correct associations. Early studies discovered null or spurious results.

More recent studies have progressed using two different research methods. While early studies on SNPs and behavior may have yielded spurious results, scientists did discover the DNA differences that lead to physical differences, like eye color. Some of these differences relate to how the brain functions and thus could drive financial decisions and other behavior. This line of research is reviewed in Chapter 5, which covers the brain in detail. The other line of research combines the individual SNPs into one genetic index, called a polygenic score. Then decision making and outcomes are evaluated against peoples' polygenic scores. This method is described in the next section.

The Genome-wide Association Method

The method for genome-wide association studies (GWAS) began by testing each of the millions of SNPs individually, with essential controls, against a specific behavior or outcome. This process identifies the gene variants that are associated with the behavior. Each variant has its own DNA location name, like ZNF169, which is the zinc finger protein #169, which helps to coordinate one or more zinc ions to stabilize the folding of the DNA sequence. A specific behavior may have several or even hundreds of gene variants that are shown to be important in determining the trait or behavior. Some gene variants will positively contribute to the behavior while others may negatively contribute to it. This process has identified the genetic variants associated with many outcomes such as the genes ZBTB7B, ACHE, RAPGEF3, RAB21, ZFHX3, ENTPD6, ZFR2, ZNF169, MC4R, and KSR2 being associated with obesity—the gene MC4R accounts for nearly 15 pounds.[11] Similarly, chromosome 15 includes the α5-α3-β4 nicotinic receptor gene cluster which is associated with smoking.[12]

The GWAS method combines the influence of all the variants into one number, which is the polygenic score for that behavior or outcome. Each person will have a polygenic score that is a factor of whichever gene variants he or she has and how they contribute to the behavior. Scholars then use this score to examine the genetic contribution to specific economic outcomes.[13] For example, 74 SNPs have been associated with educational attainment.[14] Scholars can then use these SNPs to create a polygenic score for the educational attainment of each person. The higher a person's educational attainment polygenic score, the more likely that person is to reach higher levels of education. Then, this genetic combination can be tested to determine if it is also associated with other outcomes, like labor market outcomes.[15]

Note that interpreting GWAS analysis has its limitations. One challenge is that it does not entirely separate nature from nurture effects. A person inherits their genetic code from their parents, but parents also shape rearing environments. Thus, an educational attainment polygenic score could reflect differences in biological traits related to educational attainment and also environmental influences that impact education. Studies using GWAS methods try to control for the environment by adding variables of parental education, parental resources, sibling economic attainment, and others. Nevertheless, the separation between nature and nurture is not as clear as one might expect using data on genetic markers.

Investing

Barth and colleagues used molecular genetic data in the Health and Retirement Study (HRS) to identify a biological basis for wealth attainment, risk preferences, and stock market participation.[16] They began by creating a polygenic score for each person in the HRS sample based on the educational attainment SNPs. The authors used a different sample to define the GWAS associations. Thus, the results of this study are effectively an out-of-sample test of the educational attainment polygenic score predictability of economic outcomes and decisions.

The HRS is a panel study that follows Americans over the age of 50. Surveys began in 1992 and were conducted every two years thereafter. The HRS collected genetic samples from people over the course of four waves starting in 2006. After restricting the sample to retired people with genetic typing who completed the surveys, the study's final sample includes 4,297 households and 15,670 household-year observations.

The study compared the educational attainment polygenic scores to the household wealth obtained from the HRS database. It reported a strong association by which the larger the polygenic score, the higher the amount of wealth attained. The magnitude was large—a one standard deviation increase in the score was associated with a 33 percent increase in wealth. When standard control variables like gender, age, etc., were included in the analysis, the one standard deviation increase in the score was still associated with an increase in wealth of more than 30 percent. Note that the educational attainment polygenic score was based on educational attainment, which might also influence the attainment of wealth. So, the study included education controls as well as income levels. Even after adding these control variables, the study reported that the educational attainment polygenic score was still important, although its influence was reduced by about two-thirds.

If education and income do not fully explain the association between genetics and wealth in retirement, then what else might? One possible answer is investment decisions. How a person invests over their lifecycle will have a large impact on wealth in retirement. A person who is willing to take financial risk is more likely to grow their investments than someone who is risk averse. The study used a survey-based labor risk aversion exercise, participation in the stock market, and business ownership to assess financial risk attitudes in the HRS data. The exercise asked a series of nearly identical questions that varied one parameter. The base question was:

> Choose between two jobs. The first would guarantee your current total family income for life. The second is possibly better paying, but the income is also less certain. There is a 50–50 chance the second job would double your total lifetime income and a 50–50 chance that it would cut it by 10 percent. Which job would you take?

The series replaces "10 percent" with the income losses of 20 percent, one-third, one-half, and 75 percent. A person's risk aversion is characterized by the smallest income loss percentage for which he would reject the gamble. The HRS data also included variables on whether each household was invested in the stock market and whether it owned a business. In testing the association between the educational attainment polygenic score and these measures of risk aversion, the study reported that:

- A negative relationship exists between educational attainment score and labor income risk aversion. Higher scores were associated with lower risk aversion, suggesting that people with more education were more willing to take risks.
- A positive, but weak, relationship occurs between the educational attainment score and business ownership.
- A positive and robust association exists between the educational attainment score and stock market participation.

In addition to risk aversion, a successful investor endeavors to avoid behavioral biases. One common error is that people have trouble forming accurate beliefs about the probability of events occurring, especially extreme outcomes. The HRS asked participants about the risks and uncertainties associated with the macroeconomy. They were asked to provide a probability for each of the following three events:

1   By this time next year, what is the percentage chance that mutual fund shares invested in blue chip stocks like those in the Dow Jones Industrial Average will be worth more than they are today?
2   What do you think are the chances that the U.S. economy will experience a major depression sometime during the next ten years or so?
3   And how about the chances that the U.S. economy will experience double-digit inflation sometime during the next ten years or so?

Participant answers were compared with the objectively correct answers. Observed during the period 1963–2010, the probability of the S&P 500 Index increasing in value over the course of a year was 74 percent. Historically, the probability of a severe economic contraction is 4.4 percent per year. So, over a ten-year period, the chance of a major depression can be calculated at 36 percent. Finally (examining the period 1958–2015), the odds of double-digit inflation were 3.4 percent per year, which implies a 29 percent chance over ten years. Educational attainment scores were assessed in relation to the responses in two ways: the absolute deviation between the participant response and the objective answer; and extreme responses (0 and 100 percent). The study reported that:

- For all three macroeconomic events, higher values of the educational attainment polygenic score were associated with smaller deviations between the response and the objective probability.
- For all three events, lower values of the score were related with providing an extreme response.

Overall, this study suggests that people may be partially biologically coded to nudge them in financial decisions through influence on risk aversion and biases in probabilistic thinking. Specifically, people with higher educational attainment polygenic scores are more likely to be less risk averse, own stocks, own a business, and avoid extreme beliefs about the probability of macroeconomic events.

Risk Aversion

Benjamin and colleagues studied the genetic architecture of economic preferences using multiple methods.[17] A subset of the twins in the Swedish Twin Registry were genotyped, which allows researchers to use the twin variance decomposition method and also genetic marker methods. Their final sample was 9,617 people. They provided an analysis of risk attitudes.

In addition to estimating the nature contribution using the variance decomposition method for twins in the study, they also used the dense SNP data to estimate the percentage of variation in risk aversion that can be jointly explained by the genotyped SNPs. This process is a derivative of GWAS. One criticism of GWAS is that only SNPs that show significance in first step regressions are included in the second step genetic score estimation. However, other SNPs could be significant in combinations instead of individually. This new estimation method is called genomic-relatedness-matrix restricted maximum likelihood (GREML) and estimates a lower bound for heritability contribution. GREML assumes that environmental factors are uncorrelated for individuals who are not in the same extended families. However, it estimates the level of family relatedness directly from the genetic data. Thus, it allows for more genetic sharing variations than just the three levels used in the twin studies of identical shared DNA, 50 percent shared, and no shared family relatedness.

To examine the genetic contribution to risk attitudes, an overall risk aversion was measured through a series of three risk attitude surveys and then combined into one risk aversion measure. The questions are common to other surveys and cover general risk, financial risk, and employment risk.[18] The general risk question was:

> How do you see yourself: Are you generally a person who is fully prepared to take risks or do you try to avoid taking risks? Please tick a box on the scale, where the value 1 means 'unwilling to take risks' and the value 10 means 'fully prepared to take risks.'

The second question framed risk in financial terms:

> Are you a person who is fully prepared to take financial risks or do you try to avoid taking financial risks? Please tick a box on the scale, where the value 1 means 'unwilling to take risks' and the value 10 means 'fully prepared to take risks.'

The employment risk measure came from a three-question set. It was similar to the labor risk set of questions mentioned earlier. The first question was:

> Imagine the following hypothetical situation. You are the sole provider for your household, and you have the choice between two equally good jobs:
>
> Job A will with certainty give you SEK 25,000 per month after taxes for the rest of your life.
>
> Job B will give you a 50–50 chance of SEK 50,000 per month after taxes for the rest of your life, and a 50–50 chance of SEK 20,000 per month after taxes for the rest of your life.
>
> Which job do you choose?

The other two questions varied the low-income amounts from SEK 20,000 to SEK 22,000 and SEK 17,000, respectively. Finally, the results from the general, financial, and employment risk attitudes were combined into one risk aversion measure.

The authors reported that using the standard twin variance decomposition method, about 30 percent of the variation in risk aversion was attributable to nature. This is similar to findings using other economic preferences. The GREML method estimated that the genetic contribution to risk decisions was about half of the twin estimate, at 15.8 percent. Again, this provides a lower bound for the impact of genetics on risk preferences. The study concluded that molecular genetic-based methods testing the heritability of economic preferences partially corroborate the twin-based estimates.

Summary

Does a person's genetic code influence their investing decisions? The previous chapter examined this question using the genetic variation in twins and concluded that one-fifth to one-half of all financial and economic decisions can be attributed to biology. This chapter continues the exploration of genetic influences using the dynamics and genetic variation of adoptees, their parents (biological and adoptive), and adoptive siblings. The adoptee literature suggests that biology does play an essential role in explaining economic and financial outcomes. Some estimates indicate that the contribution of nurture and unique experiences combined are twice as influential as nature. This means that biology can explain about one-third of the variation of economic outcomes between people. There is also an important role for post-birth effects (i.e., nurture and unique experiences). In this regard, gender plays a role. The economic outcomes of mothers have more influence on the outcomes of daughters, while the outcomes of fathers have more influence on sons. Thus, there is a role model aspect of nurture.

The popularity of genotyping is increasing. Since the mapping of the human genome, data samples are becoming available that include genetic markers. However, due to the magnitude of the human genome, it has been more difficult than expected to find genetic markers that drive economic decisions and outcomes. Therefore, social scientists have developed methods to test the DNA differences between people through the influence of groups of SNPs. A person's specific combination of associated SNPs are denoted as their polygenic score. An important finding is that the score associated with educational attainment also has an influence on economic outcomes and investment characteristics like risk aversion, stock market participation, and probabilistic thinking. These are the first steps in determining the mechanisms for how genetic markers may influence decisions.

Questions

1   How does adoption research examine the influence of genetics on behavior?
2   How does adoption research explain the impact of genetics in financial decision making?
3   What is the biggest issue with research using DNA data and how do analysts work around this issue?
4   What does research using the human genome say about investors' financial decisions?

Notes

1   Matthew J. Lindquist, Joeri Sol, and Mirjam Van Praag. 2015. Why do entrepreneurial parents have entrepreneurial children? *Journal of Labor Economics* 33:2, 269–296.
2   David Cesarini, Magnus Johannesson, and Sven Oskarsson. 2014. Pre-birth factors, post-birth factors, and voting: Evidence from Swedish adoption data. *American Political Science Review* November: 1–17.
3   Mikael Lindahl, Evelina Lundberg, Mårten Palme, and Emilia Simeonova. 2016. Parental influences on health and longevity: Lessons from a large sample of adoptees. NBER Working Paper No. 21946.
4   Randi Hjalmarsson and Matthew J. Lundquist. 2013. The origins of intergenerational associations in crime: Lessons from Swedish adoption data. *Labour Economics* 20:January, 68–81.
5   Erik Plug and Wim Vijverberg. 2003. Schooling, family background, and adoption: Is it nature or is it nurture? *Journal of Political Economy* 111:3, 611–641.
6   Bruce Sacerdote. 2007. How large are the effects from changes in family environment? A study of Korean American adoptees. *The Quarterly Journal of Economics* 122:1, 119–157.

7   Anders Björklund, Mikael Lindahl, and Erik Plug. 2006. The origins of international associations: Lessons from Swedish adoption data. *Quarterly Journal of Economics* 121:3, 999–1028.

8   Anders Björklund, Markus Jäntti, and Gary Solon. 2007. Nature and nurture in the intergenerational transmission of socioeconomic status: Evidence from Swedish children and their biological and rearing parents. *The B.E. Journal of Economic Analysis & Policy* 7:2, 1–23.

9   Sandra E. Black, Paul J. Devereux, Petter Lundborg, and Kaveh Majlesi. 2017. On the origins of risk-taking in financial markets. *Journal of Finance* 72:5, 2229–2278.

10 Andreas Fagereng, Magne Mogstad, and Marte Rønning. 2019. Why do wealthy parents have wealthy children? Working Paper, University of Chicago, Department of Economics, July.

11 Valérie Turcot, Yingchang Lu, Heather M. Highland, Claudia Schurmann, Anne E. Justice, Rebecca S. Fine, Jonathan P. Bradfield, et al. 2018. Protein-altering variants associated with body mass index implicate pathways that control energy intake and expenditure in obesity. *Nature Genetics* 50:1, 26–41.

12 Laura Jean Bierut. 2010. Convergence of genetic findings for nicotine dependence and smoking related diseases with chromosome 15q24-25. *Trends in Pharmacological Sciences* 31:1, 46–51, and Thorgeir E. Thorgeirsson, Daniel F. Gudbjartsson, Ida Surakka, Jacqueline M. Vink, Najaf Amin, Frank Geller, Patrick Sulem, Thorunn Rafnar, Tõnu Esko, Stefan Walter, et al. 2010. Sequence variants at CHRNB3-CHRNA6 and CYP2A6 affect smoking behavior. *Nature Genetics* 42:5, 448–453.

13 Jonathan P. Beauchamp, David Cesarini, Magnus Johannesson, Matthijs J. H. M. van der Loos, Philipp D. Koellinger, Patrick J. F. Groenen, James H. Fowler, J. Niels Rosenquist, A. Roy Thurik, and Nicholas A. Christakis. 2011. Molecular genetics and economics. *Journal of Economic Perspectives* 25:4, 57–82.

14 Aysu Okbay, Jonathan P. Beauchamp, Mark Alan Fontana, James J. Lee, Tune H. Pers, Cornelius A. Rietveld, Patrick Turley, Guo-Bo Chen, Valur Emilsson, S. Fleur W. Meddens, et al. 2016. Genome-wide association study identifies 74 loci associated with educational attainment. *Nature* 533:7604, 539–542.

15 Nicholas W. Papageorge and Kevin Thom. 2018. Genes, education and labor outcomes: Evidence from the Health and Retirement Study. NBER Working Paper No. 25114, September.

16 Daniel Barth, Nicholas W. Papageorge, and Kevin Thom. Forthcoming. Genetic endowments and wealth inequality. *Journal of Political Economy*.

17 Daniel J. Benjamin, David Cesarini, Matthijs J. H. M. van der Loos, Christopher T. Dawes, Philipp D. Koellinger, Patrik K. E. Magnusson, Christopher F. Chabris, Dalton Conley, David Laibson, Magnus Johannesson, and Peter M. Visscher. 2012. The genetic architecture of economic and political preferences. *PNAS* 109:21, 8026–8031.

18 Daniel J. Benjamin, David Cesarini, Matthijs J. H. M. van der Loos, Christopher T. Dawes, Philipp D. Koellinger, Patrik K. E. Magnusson, Christopher F. Chabris, Dalton Conley, David Laibson, Magnus Johannesson, and Peter M. Visscher. 2012. Supporting information: The genetic architecture of economic and political preferences. Available at www.pnas.org/content/suppl/2012/05/03/1120666109.DCSupplemental.

4 Do Men and Women Invest Differently?

Women have been consistently shown to take less risk than men in social settings, such as alcohol/drug use, sexual activities, driving, fighting, and gambling, among others.[1] While taking fewer risks in these social situations is likely a positive characteristic, financial risk is a different matter. Finance theory explains that some risks earn a reward, called a risk premium. Investing in the stock market is risky but earns a commensurate risk premium. Having a savings account at a bank is not risky and does not earn a risk premium. So, over time stock market returns are higher than bank account returns. If women consistently take less financial risk over their lifecycles than men, all else being equal, then women will accumulate less wealth. Thus, investing differences between men and women can have very important consequences.

Gender research examines the difference in behavior between males and females, commonly known as sexual dimorphism. The terms "gender" and "sex" are commonly interchanged; however, there are distinct differences in the research of each. Sex refers to biological differences between males and females such as women having 46 chromosomes with two X chromosomes while men have 46 chromosomes with an X and Y chromosome. Other examples include the biological differences in testosterone, estrogen, and progesterone levels. Conversely, gender typically denotes the different social or cultural norms, roles, and relationships of males and females. These gender norms are often learned behavior from the environment. Regardless of the determinants of sex/gender differences, this chapter illustrates that men and women often display different financial behaviors.

Determinants of Gender Differences

Researchers of gender examine whether differences are rooted in biology or if they sprout from social and cultural interactions, similar to the nature versus nurture debate. Regardless of any behavior that is learned as being socially or culturally accepted, there are biological differences that will always make men and women react in different ways.

Gender Biology

One of the more obvious differences in gender revolves around the production of different hormones. For example, testosterone levels are typically about four times higher in men than in women. Testosterone is a critical hormone related to risk-taking behavior (see Chapter 6), which may explain some risk-taking gender differences. Additionally, women go through a 28-day menstrual cycle which Dreher and colleagues argue modulates the reward-related neural function in women causing variations in behavior during the cycle.[2] Furthermore, Ichino and Moretti found that women are at a higher risk of absenteeism (absence from work due to illness) than men, which they argue can be attributed to complications associated with their menstrual cycle. They argued that this helps to explain a portion of the gender earnings gap.[3]

There are also neural differences between men and women. Neuro researchers have known for a long time that men and women have differences in function in the hypothalamus as it regulates sex hormones and sexual behavior.[4] However, more recent research has started to examine differences in other areas of the brain. For example, Ingalhalikar and colleagues used a large sample of 428 males and 521 females and found that women's brains have significantly stronger patterns of interconnectivity across different neural regions, including across the left and right hemisphere, than men. They conclude that different parts of the brain work together better for women than for men.[5]

Societal Differences

Unlike gender biological differences, learned societal behavior is a fluid impact that depends on many environmental factors and thus can change over time. For example, high heels were originally designed for men in the sixteenth century and pink was a boy's color in the early 1900s. Social norms for men and women change. There are significant differences between the economic and home gender roles for the millennial generation, the baby boomers, and the GI generation. However, these gender societal norms can have an enormous impact on behavior. For example, one social norm has been for men to control a family's finances, which was a possible reason for women tending to have lower financial literacy than men.[6] This difference in financial knowledge may help to explain why men and women make different financial decisions. Similarly, Riquelme and Rios found that gender-based societal norms are a contributing factor for women to adopt mobile banking. They like it because of the image it portrays and the approval it brings from other users.[7]

Additionally, women are underrepresented in STEM (Science, Technology, Engineering, and Math) careers and in the financial industry. The two main theories for this phenomenon are occupational self-selection due to preferences or abilities, and possible discrimination in the workplace.

Adams and colleagues explored reasons why only about 18 percent of people with a Chartered Financial Analyst (CFA) license are women and came to the following conclusions:[8]

- Female CFA members are less tradition-oriented and care less for family or religious customs than male CFA members and the general population of women.
- Female CFA members are less conformity oriented and care less for behaving according to social norms than male CFA members and the general population of women.
- Female CFA members are more achievement oriented, care more about their success and achievements than male CFA members and the general population of women.

The authors argued that women face more household obligations outside of work than men due to social norms, which makes it harder for women to work the long and inflexible hours associated with financial careers, thus creating a possible barrier. Similarly, Green and colleagues examined the impact of gender and job performance on Wall Street.[9] Despite the decline in social stigma about female careers from 1995 to 2005, the number of women holding analyst positions on Wall Street dropped by 2 percent. The authors also found that on average women tended to cover fewer stocks and to have less accurate earnings estimates than men. However, despite the lower performance on these measures, women were more likely than men to be designated as All-Stars, thereby indicating superior performance overall.

Financial Experiment Research

First, consider the betting behavior in the popular game show, *Jeopardy*. Jetter and Walker examined gender differences among children, teenagers, and college students from the popular TV show.[10] The authors found that female teenagers and college students wagered about 7 percent less than males during the daily double. This evidence supports the idea that women take less risk than men even though there was no difference in performance, i.e., providing the correct answer.

Powell and Ansic conducted two different experiments on the impact of gender on risk taking.[11] In the first experiment, the participants made 12 separate insurance decisions. Each decision put the participant in a situation whereby he or she had a certain amount of wealth (low or high), information about the cost of insuring an asset (low or high), and the probability of making a loss on the asset (low or high). The authors found that women were more likely to purchase insurance than men, thus demonstrating their risk aversion. In the second experiment, the subjects were given 100 euros that they could convert to U.S. dollars so as to earn interest on their dollars. However, the euro-dollar relationship could change, which would change the final return. If the dollar appreciated, the return would be higher, while a depreciation would make the return lower. To avoid the risk of converting U.S. dollars back to euros at the end of the game, subjects could choose when to convert, if ever. The results showed that women invested less money into U.S. dollars and when they did, kept it in the U.S. market for shorter time periods, thereby exhibiting higher risk aversion.

Brooks and Zank examined loss aversion and sensation-seeking behavior among men and women.[12] In this experiment, 49 students participated in 96 different trials in which they could select one ball out of a bag of 12 colored balls. One color paid a reward, while the other color had a cost. For each trial, they were told how many balls in the bag were of each color and were told of the payout if they selected the right color ball or the loss if they picked the wrong color. The gambles ranged from gaining $15 to losing $10. The first 48 trials reflected loss-averse choices where the value of the gain outweighed the value of the loss, such as a 50–50 chance of gaining $9 or losing $1. Tasks 49–96 consisted of equal probabilities of the same gain or loss or the loss outweighing the gain (i.e., a 50–50 chance of gaining $5 and losing $10). The scholars were interested in which trials the subjects chose to pick a ball and in which trials they abstained. The first 48 trials measured risk aversion while the second 48 trials measured sensation seeking. The authors found that 73 percent of women were loss averse with 9 percent of the women being classified as sensation seekers and the balance being unclassified. Meanwhile, 44 percent of men were classified as loss averse and 37 percent of them were classified as sensation seekers. This demonstrated a significant gender difference for both loss aversion and sensation seeking.

Charness and Gneezy reviewed 15 experiments from different countries to examine the impact of gender differences in the same investment task.[13] In this investment task, the subject received a certain amount of money at the outset. The person then decided how much to invest in the risky asset with known probabilities of profit and loss versus how much to keep in a risk-free investment. The authors found consistent evidence that women tended to invest lower amounts in risky investments compared to men, suggesting that women were more financially risk averse. These studies illustrate that women are more financially risk averse than men in several contexts. This may have potential consequences for wealth accumulation that could impact retirement lifestyle.

Retirement Accounts

Women tend to get paid less than men—even for the same work. Due to this earnings gap, women earn less money than men over their lifetimes. Add the fact that women don't take as much risk with their investments and the consequence is that women have significantly less savings for retirement. This ramification is further magnified in importance because women also tend to live longer than men, making it necessary for them to spread smaller retirement accounts across more years. Therefore, it is important to examine whether investment decisions are actually different between men and women.

One simple way to examine financial risk taking is to examine the ratio of equity versus debt investment in one's portfolio after controlling for other important variables such as wealth or age. Bernasek and Shwiff examined the impact of gender, risk, and retirement using survey data from faculty employed at five different universities throughout Colorado.[14] The authors found that women invested about 12 percent less of their retirement portfolio in equities compared to men. Jianakoplos and Bernasek examined the influence of gender on risk aversion using data from the 1989 Survey of Consumer Finances.[15] This dataset contains financial data for over 3,000 households. The authors found that single women are less likely to invest in risky assets like stocks compared to single men and married couples. They argued that this behavior may explain why women tend to have lower levels of wealth compared to men. Similarly, Agnew and colleagues examined the influence of gender and investing behavior using data from nearly 7,000 401(k) plans in the mid- to late 1990s.[16] The authors found that men allocate approximately 42 percent of their 401(k) retirement plan portfolio in equities, while women only allocate about 33 percent. This confirmed similar results from Sunden and Surette, who examined gender differences in defined contribution retirement plans from 1992 to 1995.[17]

Fisher and Yao also use the Survey of Consumer Finances, employing the more recent 2014 data.[18] One of the questions within this survey measures financial risk tolerance:

> Which of the statements on this page comes closest to the amount of financial risk you and your (spouse/partner) are willing to take when you save or make investments?

1   Take substantial financial risks expecting to earn substantial returns.
2   Take above average financial risk expecting to earn above average returns.
3   Take average financial risk expecting to earn average returns.
4   Not willing to take any financial risks.

Fisher and Yao found that women are significantly more likely to select average risk or no risk compared to men, which explains why women may select fewer equities in their retirement portfolios.

Trading Behavior

The previously mentioned research examined risk-averse behavior among random participants in experimental settings and a large number of retirement accounts. One may argue that these people may not be financial experts, and their risk-averse behavior may be a consequence of their lack of financial knowledge. Thus, it is useful to examine the trading behavior of professional and retail investors.

Beckmann and Menkhoff surveyed 649 professional fund managers in the United States, Germany, Italy, and Thailand with a sample of 125 women and 524 men.[19] To assess the fund managers' risk aversion, the survey asked them to answer the following questions:

(A) "In respect of professional investment decisions, I mostly act...", answered with a six-point Likert scale ranging from very risk averse to little risk averse.

(B) "In case of loss positions in my portfolio I generally wait for a price rebound instead of selling those securities," answered with a six-point Likert scale ranging from complete approval to complete contradiction.

(C) "I prefer to take profits instead of cutting losses, when I am confronted with unexpected liquidity demands," answered with a six-point Likert scale ranging from complete approval to complete contradiction.

The authors found consistent evidence that these women in the financial industry preferred less risk (question A), were more loss averse (question B), and were more likely to commit the disposition effect (question C) in Germany, Italy, and Thailand, but not in the United States. The authors also found that this difference in risk aversion was not impacted by investor overconfidence. The authors went on to examine risk taking in a tournament setting (which occurs frequently in the financial industry). The survey gave the fund managers the scenario of either outperforming the benchmark or underperforming the benchmark. Based on their performance, the fund managers could either stick to their previous investment strategy or change their strategy to either take more or less risk if they underperformed. If they overperformed, the managers could either "lock in" their good performance, or take even more risks to try to become the manager with the highest returns. First, the study found that female fund managers in Italy and Thailand were more likely to change their strategy and take less risk if they overperformed the benchmark, which is consistent with risk-averse behavior. Similarly, female fund managers in the United States tended to reduce risk in their portfolios after underperformance. However, the authors found that female fund managers in Germany and Thailand were more likely to change their strategy and take more risks if their fund underperformed, demonstrating that there are still gender differences in risk taking that are not entirely understood.

One limitation of the previous study is that it is possible that people's actual trading behavior may deviate from what they say they would do when completing a survey. As such, Dwyer and colleagues examined the investment decisions of nearly 2,000 retail investors who held mutual funds in their portfolio. What was the largest mutual fund holding of the investor? Was it a low risk investment in a money market fund? Or was it a high-risk investment like a stock mutual fund? Or something in between? The results of the study are illustrated in Figure 4.1.

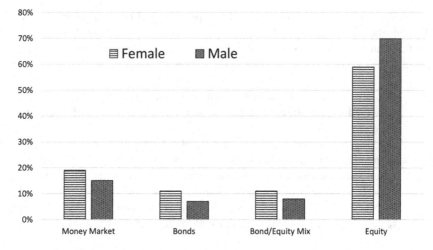

Figure 4.1 Gender and Mutual Fund Choice

The graph shows that 30 percent of women had their largest holding in a low-risk investment like the money market or a bond fund. This compares with only 22 percent of men allocating their largest holding to a low-risk investment. Conversely, 70 percent of men held equities as their largest mutual fund compared to only 59 percent of women, thus demonstrating consistent risk aversion among female investors. The authors also found that women were more likely than men to hold money market or bond mutual funds as the riskiest asset they owned.

In one of the most well-known studies on gender differences in financial behavior, Barber and Odean examined the impact of gender on trading decisions from more than 35,000 retail investor portfolios.[20] The authors first examined investment risk by examining the portfolios' volatility and beta. They reported that men's portfolios had higher volatility and had a higher beta than women's portfolios. Furthermore, the authors hypothesize that men will trade more frequently than women. High trading is a ramification of the overconfidence behavioral bias. Excessive trading can be a detriment to portfolio returns due to extra trading costs. The authors found that women have a monthly portfolio turnover of 4.4 percent, demonstrating that they tend to sell off 4.4 percent of their portfolio each month to buy different assets. Conversely, the authors found that men typically have a monthly portfolio turnover of 6.4 percent, which suggests 45 percent more trading activity. Thus, men appeared to exhibit more of the overconfidence bias than women. Additionally, when examining only single females and single males, the authors found that male portfolio monthly turnover exceeded female turnover by 3 percentage points. The authors also reported that men tended to have lower portfolio returns compared to women because of the higher portfolio turnover. Finally, the portfolio return differences were not caused by women's lower risk tolerance or willingness to take gambles.

In another study using retail investors, Grinblatt and Keloharju examined the impact of sensation seeking and overconfidence using Finnish investors' portfolio trading histories.[21] The authors first found that Finnish men trade more than Finnish women, confirming the results of Barber and Odean with a different sample from another country. Grinblatt and Keloharju posited that this increased level of trading could be caused either by overconfidence, as suggested by Barber and Odean, or by sensation seeking, which is another trait that men typically exhibit to a greater extent than women. The authors argued that one frequent behavior that reveals a person's sensation-seeking behavior can be found in their driving history. Research demonstrates that men overwhelmingly receive more speeding tickets than women. As such, the authors matched the Finnish portfolio data with people's driving histories to examine if sensation seeking may explain the gender differences in portfolio turnover. In order to compare the explanation of sensation seeking with overconfidence, the authors used data from the Finnish armed forces psychological profile that asked questions about self-confidence and ability. The authors found that each speeding ticket that an investor received was associated with an additional 10 percent in trading. In addition, using the military confidence score, the authors also found that overconfidence, in addition to sensation seeking, played a critical role in gender difference in investor behavior. In summary, compared to men, women were shown to trade less, and to exhibit lower overconfidence and sensation-seeking behavior.

Gender in Corporate Finance

While there appears to be behavioral differences between male and female investors, how does gender diversity affect a corporation? Carter and colleagues examined the impact of gender diversity on boards of directors in Fortune 1000 firms.[22] The authors found that having women serve on a board of directors increased the firm's overall value. Similarly, Campell and Minguiez-Vera examined the issue of gender diversity in the boardroom using data from Spanish companies.[23] These authors also found that a higher presence of women serving on a company's board of directors increased the firm's value. Furthermore, Bear and colleagues found a positive relationship between the number of women on the board of directors and the company's corporate social responsibility score.[24]

Barua and colleagues examined whether male and female chief financial officers (CFOs) treat the company's accounting accruals differently.[25] The authors hypothesized that because women are typically more risk averse, more ethical, and less overconfident, their firm will have a higher accrual quality than a male CFO's firm. This means that female CFOs will be less prone to making use of earnings management techniques to artificially smooth the earnings per share in an effort to inflate the stock price.

Consistent with their hypothesis, the authors found that having a female CFO leads to higher accrual quality and lower accrual estimation errors.

Following the same idea, Francis and colleagues examined financial reporting characteristics by examining how the company changed its reporting decision making when the firm changed from having a male CFO to a female CFO.[26] The authors found that female CFOs exhibited more conservative financial reporting. Furthermore, firms with a female CFO had lower litigation risk, default risk, systematic risk, equity-based compensation, and dividend payouts while having higher tangibility levels.

Nature or Social Norms?

So far, this chapter has demonstrated significant gender differences in risk taking and financial behavior. However, the studies offered very little explanation as to the fundamental source of differences. Are the behavior differences driven by differences in biology or social norms and environment?

Feminism and Masculinity

To explore the nature versus social norms question, Cross and colleagues conducted a meta-analysis of the gender differences found in sensation-seeking behavior using Zuckerman's Sensation Seeking Scale.[27] This survey examined four components of sensation seeking: thrill and adventure seeking; disinhibition; boredom susceptibility; and experience seeking. Historically, men have higher scores on all measures except experience seeking. Note that the biological differences between men and women do not change over time. However, social norms do change. Thus, the study examines sensation-seeking behavior over time. If the gender-based behavior differences change over time, then social norms may drive those differences. Alternatively, if the differences in behavior do not change over time, then biology may drive the behavior. The study found that the gender differences for these traits were stable across decades except for the thrill and adventure category, for which the gender difference has significantly decreased. Overall, these results show a strong role for biology driving the differences between the financial decisions of men and women. The authors argued that the thrill and adventure category finding could have been caused by social norms changes, more women participating in athletics, or outdated survey questions.

The literature suggests that women have stronger emotional reactions to experiences than do men. So, Eriksson and Simpson examined the impact of emotional reactions as an explanation for gender-based behavior differences in risk taking involving gambling.[28] The authors asked subjects the following questions:

1   Suppose you have to pay 35 dollars to enter a lottery where there is one chance in nineteen (1/19) of winning 1,000 dollars. Imagine that you

entered the lottery, at a cost of 35 dollars and lost. How would you feel about the loss on a scale of 0 (neutral) to 5 (extremely bad)?

2   Imagine that you entered the same lottery, at a cost of 35 dollars, and won 1000 dollars. How would you feel about the win on a scale from 0 (neutral) to 5 (extremely good)?

3   Assuming that you were assured that this was a completely fair lottery (with one chance in nineteen of winning 1,000 dollars), would you be willing to pay 35 dollars to enter?

The authors found that women had a stronger emotional response to the scenarios. Specifically, women were more likely to give higher scores on both questions as being more upset if they lost and being happier if they won. Furthermore, the authors found that women were also less likely to participate in the gamble and argued that this was due to the emotions of either winning or losing the gamble.

Consider that sex is a biological term that refers to a person being either male or female. However, from a social viewpoint, some consider gender to be a continuum between the feminine and masculine. Bem's Sex Role Inventory attempts to categorize people on a gender scale. Some men may measure as more feminine while some women may measure as more masculine. This is called gender expression. Is it the dichotomy of sex or the scale of gender that explains financial decision-making differences between men and women? Meier-Pesti and Penz examined these measures of sex and gender on financial risk taking.[29] The subjects were asked questions about risk taking in their investment portfolios and about future investments. When using the biological sex dichotomy of men and women, the authors found that men take significantly higher financial risks. However, after accounting for gender expression, there were no risk-taking differences, suggesting that the difference in financial risk taking may be due to societal norms.

Niederle and Vesterlund examined competitive behavior among men and women.[30] The subjects were asked to add as many sets of five two-digit numbers without a calculator (i.e., 21 + 35 + 48 + 29 + 83) as they could during different tasks. In task 1, they were given five minutes to complete the problems and were paid according to the number of correct answers. In task 2, a tournament scenario was used in which only the person who answered the most answers correctly was paid. Note that the first two tasks create a baseline for math ability. In task 3, the participants were asked if they wanted to do the task under no competition (i.e., task 1) or competition (i.e., task 2). This task asks if the participant wants the risk of competition or not. Finally, in task 4, the participants could select one of their previous performances to be put into the tournament setting (i.e., they did not retake the task, but used a past performance). The authors found no gender differences in the number of correct answers in tasks 1 and 2. Despite no gender differences in ability, for task 3 women chose to participate in the tournament only 35 percent of the time while men participated 73 percent of the time, demonstrating strong evidence that women shy away from competition. Why do women tend to shy away from competition? The study reports the following evidence:

- Men were more overconfident than women and believed they earned more correct answers during the task, which accounted for about 27 percent of the gap.
- Women were more risk averse than men, which accounted for 30 percent of the gap.
- The remaining 43 percent of the gender gap was attributed to women's desire to shy away from competition.

Financial Literacy

Differences in financial risk taking between men and women could also be a ramification of difference in financial literacy. Intuitively, people who are less informed about financial matters would choose to avoid unfamiliar risky situations. A low level of financial literacy is a major problem throughout the world. The National Financial Capability Study examined the financial literacy of American citizens by asking the following five simple financial questions:[31]

1   Suppose you have $100 in a savings account earning 2 percent interest a year. After five years, how much would you have?

    a   More than $102
    b   Exactly $102
    c   Less than $102

2   Imagine that the interest rate on your savings account is 1 percent a year and inflation is 2 percent a year. After one year, would the money in the account buy more than it does today, exactly the same, or less than today?

    a   More
    b   Same
    c   Less

3   If interest rates rise, what will typically happen to bond prices? Rise, fall, stay the same, or is there no relationship?

    a   Rise
    b   Fall
    c   Stay the same
    d   No relationship

4   A 15-year mortgage typically requires higher monthly payments than a 30-year mortgage but the total interest over the life of the loan will be less.

a   True

b   False

5   Buying a single company's stock usually provides a safer return than a stock mutual fund or ETF.

a   True

b   False

The study scored the test as a pass if the person could answer four or five questions correctly and a fail if three or fewer answers were correct. In 2009, 58 percent of Americans scored a fail. Unfortunately, the percentage of failures increased to 61 percent in 2012, 63 percent in 2015, and 66 percent in 2018, demonstrating a problem that may be getting worse. Furthermore, this study found that in 2018 men scored one-half question better than women, demonstrating a large gender difference in financial literacy.

Almenberg and Dreber examined the impact of financial literacy and gender for stock market participation using data from the consumer survey provided by the Swedish Financial Supervisory Authority.[32] The authors found that lower financial literacy scores by women can partially explain the gender gap in stock market participation. The difference in risk-taking preferences continue to explain part of the difference in financial behavior.

Similarly, Halko and colleagues examined gender differences in risky portfolio holdings in Finland.[33] The authors collected information from investment advisors, finance students, and investors on their risk attitudes, investment knowledge, and demographic information. The study found that after controlling for risk aversion scores and investment knowledge, there was no difference in asset allocation choices by men and women. This suggests that the differences between men and women can be attributed to differences in risk aversion and financial literacy.

Bannier and Schwarz examined the impact of actual and perceived financial literacy on stock market participation using data from a German financial behavior survey.[34] The study found that men scored higher on financial literacy which explained part of the gender gap in market participation and future retirement planning. Furthermore, the authors found that men were overconfident in their financial knowledge, while women were less confident in their financial knowledge than they should have been. They argue that the gender difference in perceived financial knowledge plays a more significant role in financial decisions than their actual amount of financial literacy.

Taylor and Wozniak conducted an experiment to see how men and women acquire financial information in an experimental context.[35] Participants were given $150 to acquire information and invest. During the task, the subjects made 60 decisions about the desirability of investing in a mutual fund (i.e., a risky asset) or in a money market account earning 2 percent (i.e., a safe asset). Prior to making a decision, the subject was given the previous year's return of the mutual fund from Morningstar. In addition to knowing the previous year's return, the subjects could purchase additional information for $0.50 each. The information choices included: (1) mutual fund name; (2) mutual fund category; (3) standard deviation of return; (4) dividend yield; (5) previous year's market return; (6) Morningstar rating; (7) manager tenure; (8) expense ratio; (9) turnover ratio; and (10) two-year relative return. Some of these information items are more important than others. By examining which pieces of information were purchased, the authors determined a subject's degree of financial literacy. The study reported that men scored higher on financial literacy and were more willing to purchase information about the mutual fund than women.

Ke examined the impact of gender norms on household financial decisions.[36] The author found that households with a financially sophisticated husband were more likely to participate in the stock market compared to households where the wife had a high level of financial knowledge. The study reported that these results were stronger when the household was raised with traditional gender norms, such as having the mother stay at home while the children were young, or growing up in a southern state. These are social norm explanations for gender financial differences.

The Gender of Others

This chapter has discussed the impact of gender on financial decisions. However, it is possible that financial decisions can also be influenced by the gender of those who are close to you. Two fascinating studies show that a man's financial decisions change if he becomes a father to a daughter.

Cronqvist and Yu examined the firm policies of male chief executive officers (CEOs) with and without daughters.[37] The authors argued that having a daughter could influence a male CEO to make decisions that are more beneficial to society as a whole rather than simply maximizing the shareholder value. As such, the authors examined the difference in firm policies related to corporate social responsibility (CSR) among CEOs with and without daughters. The authors found that if the CEO had a daughter, the firm's CSR score was roughly 12.5 percent higher, thus supporting their hypothesis. This CEO-daughter effect was strongest for CSR diversity ratings, but also increased other pro-social policies such as environmental concerns and employee relationships.

Pogrebna and colleagues examined the impact of having a female baby on parental financial risk taking by doing an experiment with expecting parents before and after they were told the sex of their soon to be born child.[38] The parents participated in a gambling task whereby they had to make a choice between a lower possible payout with a higher probability versus a higher possible payout with a lower probability. The results of the study are presented in Figure 4.2.

The graph shows that risk aversion scores were roughly 4 before the parents knew the gender of their child. After finding out that they were

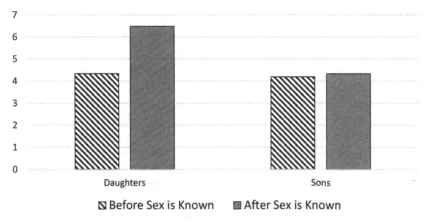

Figure 4.2 Risk Aversion Scores of Parents

having a son, the scores slightly increased to about 4.25. However, after finding out that they were having a daughter, the risk aversion scores jumped to about 6.5. These results show that the impact of having a female child starts influencing parents' financial decisions from the moment the gender of their baby is known.

Furthermore, Booth and Nolen examined the impact of the environment on people's risk-taking behavior.[39] The authors hypothesized that because of social norms, women would be less likely to take risks when men are around them. To test this theory the authors conducted an experiment by using students from eight public schools in the United Kingdom, with an average age of under 15 years old. Four of the schools were single-sex and the others were co-ed. The students were put into groups of four that were either all girls, all boys, or mixed gender. Once in their groups, the students were asked to solve as many paper mazes as they could. The students could select from options on how they got paid. For example, one option was to earn 50 cents per correctly completed maze. The other option was a gamble to get $2 per maze correctly solved if the student solved more mazes than the other three students in the group. The study reports the following results:

- Girls from co-ed schools selected the gamble 36 percent less often than boys.
- Girls from single-sex schools were as likely to select the gamble as boys.
- Girls were more likely to select the gamble when they were in a group of all girls.
- Boys from single-sex schools selected the gamble just as often as boys from co-ed schools regardless of the gender mix of their group.

The authors argued that these results were due to social norm pressure on females to conform to gender stereotypes.

Summary

Gender appears to have a significant impact on financial decisions as there is strong evidence from multiple settings that women take less risk than men. This difference in risk taking is not just among the everyday person, but also among investors. Female retail investors hold less equity and trade less frequently than male investors. One argument is that women tend to take less risk in their portfolios because they are less overconfident and less likely to exhibit sensation-seeking behavior. However, other arguments are that the differences in financial behavior are caused by emotions, competitive behavior, or financial literacy differences. In addition, social norms play an important role in shaping a person's risk preferences. Furthermore, the gender of those who are close to you can also influence a person's behavior, as demonstrated by fathers with daughters making less risky decisions compared to fathers with sons. Overall, this chapter has demonstrated significant evidence of differences between the financial decisions made by men and women, with some differences being a benefit to women, and others being a hindrance.

Questions

1 What is the difference between biological sex differences and gender societal norms? How does this relate to nature versus nurture?
2 What does experimental research report when examining differences in risk aversion between men and women?
3 What differences between men and women are found retirement accounts as regards the decision made?
4 How do men and women invest differently? What explanations do researchers have for these findings?
5 What gender differences are there between male and female business executives?
6 How does financial sophistication vary among men and women?

Notes

1 James P. Byrnes, David C. Miller, and William D. Schafer. 1999. Gender differences in risk taking: A meta-analysis. *Psychological Bulletin* 125:3, 367–383.
2 Jean-Claude Dreher, Peter J. Schmidt, Philip Kohn, Daniella Furman, David Rubinow, and Karen Faith Berman. 2007. Menstrual cycle phase modulates reward-related neural function in women. *Proceedings of the National Academy of Sciences* 104:7, 2465–2470.
3 Andrea Ichino and Enrico Moretti. 2009. Biological gender differences, absenteeism, and the earnings gap. *American Economic Journal: Applied Economics* 1:1, 183–218.
4 Seymour Levine. 1966. Sex differences in the brain. *Scientific American* 214:4, 84–92.
5 Madhura Ingalhalikar, Alex Smith, Drew Parker, Theodore D. Satterthwaite, Mark A. Elliott, Kosha Ruparel, Hakon Hakonarson, Raquel E. Gur, Ruben C.

Gur, and Ragini Verma. 2014. Sex differences in the structural connectome of the human brain. *Proceedings of the National Academy of Sciences* 111:2, 823–828.

6 Andrea Hasler and Annamaria Lusardi. 2017. The gender gap in financial literacy: A global perspective. Global Financial Literacy Excellence Center, The George Washington University School of Business, July.

7 Hernan E. Riquelme and Rosa E. Rios. 2010. The moderating effect of gender in the adoption of mobile banking. *International Journal of Bank Marketing* 28:5, 328–341.

8 Renee B. Adams, Brad M. Barber, and Terrance Odean. 2016. Family, values, and women in finance. Available at SSRN 2827952.

9 Clifton Green, Narasimhan Jegadeesh, and Yue Tang. 2009. Gender and job performance: Evidence from Wall Street. *Financial Analysts Journal* 65:6, 65–78.

10 Michael Jetter and Jay K. Walker. 2017. Gender differences in competitiveness and risk-taking among children, teenagers, and college students: Evidence from Jeopardy! Available at https://ssrn.com/abstract=3092525.

11 Melanie Powell and David Ansic. 1997. Gender differences in risk behaviour in financial decision-making: An experimental analysis. *Journal of Economic Psychology* 18:6, 605–628.

12 Peter Brooks and Horst Zank. 2005. Loss averse behavior. *Journal of Risk and Uncertainty* 31:3, 301–325.

13 Gary Charness and Uri Gneezy. 2012. Strong evidence for gender differences in risk taking. *Journal of Economic Behavior & Organization* 83:1, 50–58.

14 Alexandra Bernasek and Stephanie Shwiff. 2001. Gender, risk, and retirement. *Journal of Economic Issues* 35:2, 345–356.

15 Nancy Ammon Jianakoplos and Alexandra Bernasek. 1998. Are women more risk averse? *Economic Inquiry* 36:4, 620–630.

16 Julie Agnew, Pierluigi Balduzzi, and Annika Sunden. 2003. Portfolio choice and trading in a large 401 (k) plan. *American Economic Review* 93:1, 193–215.

17 A. E. Sunden and B. J. Surette. 1998. Gender differences in the allocation of assets in retirement savings plans. *American Economic Review* 88:2, 207–211.

18 Patti J. Fisher and Rui Yao. 2017. Gender differences in financial risk tolerance. *Journal of Economic Psychology* 61:August, 191–202.

19 Daniela Beckmann and Lukas Menkhoff. 2008. Will women be women? Analyzing the gender difference among financial experts. *Kyklos* 61:3, 364–384.

20 Brad M. Barber and Terrance Odean. 2001. Boys will be boys: Gender, overconfidence, and common stock investment. *Quarterly Journal of Economics* 116:1, 261–292.

21 Mark Grinblatt and Matti Keloharju. 2009. Sensation seeking, overconfidence, and trading activity. *Journal of Finance* 64:2, 549–578.

22 David A. Carter, Betty J. Simkins, and W. Gary Simpson. 2003. Corporate governance, board diversity, and firm value. *Financial Review* 38:1, 33–53.

23 Kevin Campbell and Antonio Mínguez-Vera. 2008. Gender diversity in the boardroom and firm financial performance. *Journal of Business Ethics* 83:3, 435–451.

24 Stephen Bear, Noushi Rahman, Corinne Post. 2010. The impact of board diversity and gender composition on corporate social responsibility and firm reputation. *Journal of Business Ethics* 9:2, 207–221.

25 Abhijit Barua, Lewis F. Davidson, Dasaratha V. Rama, and Sheela Thiruvadi. 2010. CFO gender and accruals quality. *Accounting Horizons* 24:1, 25–39.

26 Bill Francis, Iftekhar Hasan, Jong Chool Park, and Qiang Wu. 2015. Gender differences in financial reporting decision making: Evidence from accounting conservatism. *Contemporary Accounting Research* 32:3, 1285–1318.

27 Catharine P. Cross, De-Laine M. Cyrenne, and Gillian R. Brown. 2013. Sex differences in sensation-seeking: A meta-analysis. *Scientific Report* 3:2486. doi:10.1038/srep02486.

28 Kimmo Eriksson and Brent Simpson. 2010. Emotional reactions to losing explain gender differences in entering a risky lottery. *Judgment and Decision Making* 5:3, 159–163.

29 Katja Meier-Pesti and Elfriede Penz. 2008. Sex or gender? Expanding the sex-based view by introducing masculinity and femininity as predictors of financial risk taking. *Journal of Economic Psychology* 29:2, 180–196.

30 Muriel Niederle and Lise Vesterlund. 2007. Do women shy away from competition? Do men compete too much? *Quarterly Journal of Economics* 122:3, 1067–1101.

31 Marco Angrisani, Arie Kapteyn, and Annamaria Lusardi. 2016. The national financial capability study: Empirical findings from the American life panel study. *FINRA Report*, November.

32 Johan Almenberg and Anna Dreber. 2015. Gender, stock market participation and financial literacy. *Economics Letters* 137:December, 140–142.

33 Marja-Liisa Halko, Markku Kaustia, and Elias Alanko. 2012. The gender effect in risky asset holdings. *Journal of Economic Behavior & Organization* 83:1, 66–81.

34 Christina E. Bannier and Milena Schwarz. 2017. Skilled but unaware of it: Occurrence and potential long-term effects of females' financial underconfidence. *Econstar*, February.

35 Matthew P. Taylor and David Wozniak. 2018. Gender differences in asset information acquisition. *Journal of Behavioral and Experimental Finance* 20:December, 19–29.

36 Da Ke. 2018. Who wears the pants? Gender identity norms and intra-household financial decision making. *Gender Identity Norms and Intra-Household Financial Decision Making*, August 4.

37 Henrik Cronqvist and Frank Yu. 2017. Shaped by their daughters: Executives, female socialization, and corporate social responsibility. *Journal of Financial Economics* 126:3, 543–562.

38 Ganna Pogrebna, Andrew J. Oswald, and David Haig. 2017. Female babies and risk-aversion. IZA Discussion Paper No. 10717, May.

39 Alison Booth and Patrick Nolen, P. 2012. Gender differences in risk behaviour: Does nurture matter? *The Economic Journal* 122:558, F56–F78.

Section II
Physiology

5 Brain Function and Financial Decisions

Every action taken and every decision made, including financial decisions, originates in the brain. Neuroscience examines the structure and function of the brain. Neuroeconomics is a subcategory of neuroscience that analyzes the structure and function of the brain during financial and economic decision making. Traditional finance assumes that all market participants are rational. That would stipulate that everyone's neurochemistry behaves in the same way. However, when one person's sensitivity to a neurotransmitter, like dopamine, is different from another person's, these people can make different decisions. For example, medical research finds that psychopaths have structural and functional differences from normal people in the amygdala and prefrontal cortex, which are the neural regions that regulate emotion and decision making, respectively. This causes psychopaths to have persistent antisocial behavior as well as impaired empathy and remorse. Therefore, psychopaths make different social decisions compared to non-psychopaths; it is also likely they may make different financial decisions. However, a person does not have to be a psychopath to make irrational financial decisions. Each person has different sensitivities and nuances in neural functions that result in different financial decisions.

The Brain

Anatomy of the Brain

The brain is typically discussed through its different regions or by its limbic system. Figure 5.1 shows the main areas of the brain. The four major lobes that make up the main areas of the brain are the frontal lobe, parietal lobe, temporal lobe, and occipital lobe. The frontal lobe is responsible for motor function, problem solving, planning, memory, impulse control, and social behavior. These functions are vital for making sound financial decisions. The parietal lobe contains the functions of sensation, perception, the interpretation of visual information, and the processing of language and arithmetic. Many behavioral biases are based on emotions, perception, and interpreting information within a decision frame. Thus, there is the potential for psychological

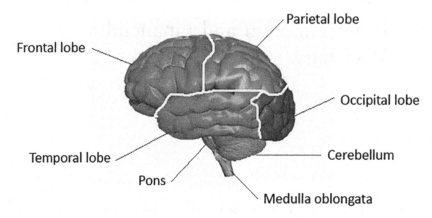

Frontal lobe

Parietal lobe

Occipital lobe

Temporal lobe

Pons

Cerebellum

Medulla oblongata

Figure 5.1 The Main Areas of the Brain

biases stemming from the parietal lobe. The temporal lobe processes hearing, memory, and language. The occipital lobe processes visual information. Additional brain structures include the cerebellum and the brain stem, which is made up of the pons, cerebral peduncles, and medulla oblongata. The cerebellum and stem structures beneath the brain are responsible for communication between the brain and the body and control the essential bodily functions like breathing, heart rate, eye movement, coordination, speech, etc.

The limbic system is the complex system of nerves and networks that are involved in motivation, emotion, learning, and memory. Figure 5.2 shows that the limbic system is made up of the thalamus, cingulate gyrus, fornix, amygdala, and hippocampus. The amygdala is important in economic decision making as it is responsible for emotions, especially fear, arousal, and general emotional stimulation. The cingulate gyrus processes conscious emotional experience and behavior regulation, while the hippocampus is associated with memory, especially long-term memory. The fornix connects parts of the limbic system. The thalamus relays sensory signals among the lobes of the brain. Table 5.1 details the function of these regions.

In addition to the major lobes and limbic system, the striatum is also essential for decision making. It is a critical component of the reward system. It consists of a small group of connecting subcortical structures located in the forebrain. The striatum is divided into dorsal and ventral sections. The dorsal striatum contains the caudate and putamen, while the ventral striatum contains the nucleus accumbens. The dorsal striatum is part of the basal ganglia that facilitates voluntary movement. The ventral striatum and nucleus accumbens have been studied for their role in pleasure, reward, reinforcement, and compulsive experiences. The ventral striatum is one of the main components involved in addiction. As such, it is activated when a person participates in or anticipates participating in activities known to result in pleasure. It is believed that the nucleus accumbens plays a

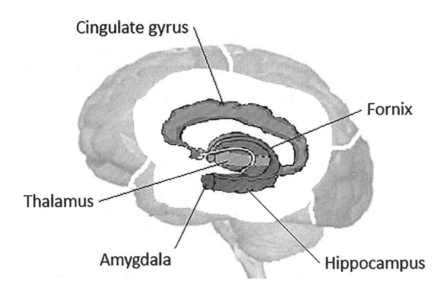

Figure 5.2 The Limbic System

Table 5.1 Functions and Locations in the Brain

Brain Region	Function
Major Lobes	
Frontal Lobe	Motor function, problem solving, planning, memory, impulse control, and social behavior
Parietal Lobe	Sensation, perception, interpreting visual information, and processing language and arithmetic
Temporal Lobe	Processes hearing, memory, and language
Occipital Lobe	Processes visual information
Stem Structure	
Pons	Respiration, cardiovascular function, eye movement, and balance
Medulla Oblongata	Vital processes such as heart rate, respiration, and blood pressure
Cerebellum	Movement
Limbic System	
Fornix	Connects the hippocampus to the other parts of the limbic system
Hippocampus	Memory
Cingulate Gyrus	Processes conscious emotional experience
Amygdala	Emotion, learning, and memory
Thalamus	Relay station for senses and the cortex of major lobes

mediating role in reward processing. That is, the nucleus accumbens weights the possible pleasure or pain experience when we decide whether to take part in a task. Finally, the insula is a small region of structures that either connect to or send signals to the limbic system. The insula's purpose is self-awareness, cognitive function, and interpersonal experience related to emotion. It is often associated with avoidance behavior.

Neurotransmitters

Neurotransmitters are chemicals released by neurons (the nerve cells) to transmit messages between neurons. Dopamine is a neurotransmitter that activates the ventral tegmental area when a person does anything that may provide pleasure or pain. These neurons project to and activate the nucleus accumbens. The nucleus accumbens is a vital component of the dopaminergic pathway. For example, if someone is offered a slice of delicious pizza, the brain releases dopamine, and the nucleus accumbens disentangles not only the benefits of eating a delicious slice of pizza but also the potential health repercussions. Additionally, dopamine can affect mood as rewards tend to make us feel good. Moreover, drugs including cocaine, nicotine, and heroin cause significant increases in dopamine which is why it is widely studied in addiction research. Therefore, the cause of the striatum and nucleus accumbens activation may be caused by the neurotransmitter dopamine, which influences many types of decisions including financial decisions.

Serotonin is a neurotransmitter that plays a role in appetite, emotion, and cognitive and autonomic functions. Additionally, it plays a crucial role in mood balance, anxiety, and happiness. Low serotonin levels are linked to depression, although it is unknown if low serotonin levels cause depression, or depression causes low serotonin levels. Moreover, drugs such as Ecstasy (MDMA) and LSD cause a significant increase in serotonin levels. Given its effect on mood, it is thought that serotonin effects financial decisions.

Types of Decision Processes

Neuroscience does more than merely locate where neural activity is situated during specific tasks. Its true purpose is to understand how the brain works by using regional activity differences to reveal brain organization and function. Shiffrin and Schneider first proposed the distinction between automatic and controlled processes.[1] Many researchers have developed similar two-system models including Lee and colleagues who called them rational and experiential processes.[2] A serial process is one that uses a step-by-step logic. This means that if asked how an individual made a choice, they could walk you through their process of decision making. For example, if someone asked how you drive a car, you would walk them through the steps of getting into the car, putting the key into the ignition, starting the car, etc.

Automatic (experiential) processes are the opposite of controlled processes because these decisions are made subconsciously and are relatively effortless. An example of this is muscle memory in athletes. If someone is playing basketball and intends to take a shot, he is carrying out many tasks at once without realizing all of them, such as looking at all the other players on the court to determine if they are open, moving the ball from a dribble to a shooting position, jumping, shooting, etc. While the athlete may be able to walk you through each of these steps, he would not be able to explain the angle of his elbow to the wrist, the number of inches he jumped off the floor, or how much force he used in his shot. All of these decisions are made subconsciously over time and with practice. Another example of controlled processes is emotion, which may cause individuals to act irrationally.

Controlled processes typically occur in the frontal lobe of the brain, while automatic processes occur in the amygdala and hypothalamus (limbic system), occipital lobe, parietal lobe, and temporal lobe of the brain. Neuroeconomics is chiefly concerned with why people behave irrationally. Studies seek to discover whether people use the controlled processes and simply make errors in their calculations or are using the automatic processes and subconsciously making irrational decisions.

The tools that are most frequently used to examine neural activity are functional magnetic resonance imaging (fMRI), positron emission tomography (PET), and the electroencephalogram (EEG). The newest and most accurate tool is fMRI. It uses a strong magnetic field and radio waves to create detailed images of the brain's activity. PET scans measure blood flow in the brain, which can be used as a proxy for neural activity. The EEG records electrical activity from the brain.

Economic Decisions and the Brain

Financial Risk and Reward

Carvalho Júnior and colleagues used an EEG to conduct an experiment that analyzed the neural activity of how auditors and accountants make judgments when assessing evidence for decisions involving the auditing task of determining whether the firm is a "going concern." This refers to a company's ability to remain in business for the foreseeable future. The results of the study show that these two groups used different parts of the brain to reach similar conclusions. Accountants used their frontal medial cortex more than auditors.[3] Thus, it is important to remember that utilizing different brain functions does not necessarily result in different decisions or conclusions.

The Balloon Analog Risk Task (BART) is an economic test in which the subject earns money every time they push a button to inflate a balloon. However, there is an unknown threshold whereby if the subject chooses to inflate the balloon one more time it will pop, and they will lose all the money they have earned in that round. Therefore, every click to inflate the balloon increases the amount of money the subject can earn, but simultaneously increases the risk that the balloon will pop. Consider that you have already clicked the balloon several times and have earned $45. Do you click it again to get to the next level, thereby risking it all? Researchers used fMRI on subjects participating in the BART and found that the ventral and dorsal striatum, anterior insula, dorsal lateral prefrontal cortex, and anterior cingulate/medial frontal cortex are responsible for assessing risk.[4] The activity of the ventral striatum demonstrates that the subjects are calculating the risk and reward of their ensuing decisions, while the anterior insula may be activated to avoid the pain of losing all their money. Additionally, this result shows that subjects use the rational part of their brain (the frontal lobe) to assess the risk of the balloon popping and thus losing their gains.

One essential economic question is how people respond to ambiguity and uncertainty. An example of ambiguity is tested when a participant in an experiment has to select between two buckets in order to earn $50 for picking a red ball and $25 for picking a black ball. In bucket A there are 50 red balls and 50 black balls, but in bucket B there is an unknown number of red and black balls. Given this scenario, most people pick from bucket A to remove the ambiguity. Using fMRI, research finds that people who pick the ambiguous bucket B use their lateral prefrontal cortex, which is also negatively related to impulsive behavior. People who pick from bucket A with the known number of balls use their posterior parietal cortex.[5] Thus, people who are more rational (frontal lobe) and less impulsive are better equipped to handle ambiguous decisions.

In another experiment about ambiguity, Hsu and colleagues conducted an experiment in which subjects had to select between two decks of 20 cards.[6] In the first deck, there were ten red cards and ten blue cards, while the number of red and blue cards in the second deck was unknown (ambiguous). The subjects had the option of guessing the color of a card drawn, which would earn them $10 if they guessed correctly or $0 if incorrect. Or they could avoid the uncertainty and select a sure gain of $3. The results show that the orbitofrontal cortex, prefrontal cortex, and amygdala are more active during ambiguous decisions and the striatum is more active when the probability of risks is known. These results imply that ambiguity lowers the anticipated reward of the gambles.

Stock selection is one of the most ambiguous decisions to be made because future returns are unknown. In fact, Antoniou and colleagues find that the ambiguity of the stock market is a significant reason why households do not participate in the stock market.[7] Therefore, it is expected that investors will use the rational part of their brain (prefrontal lobe) and emotion (amygdala) in stock selection. Of course, those who use the prefrontal lobe and those who use their amygdala are likely to choose different stocks.

Since 2008 one of the most researched topics in finance has concerned market bubbles. Smith and colleagues conducted an experiment in which subjects could buy or sell any of six assets in each of the 50 trading periods in which they made decisions.[8] During each period, the subjects could select the stock with a dividend that had a 50–50 chance of earning either $0.40 or $1.00 or they could earn 5 percent on non-invested money. Given that there was a 50–50 chance of $1 or $0.40 the expected dividend was $0.70. Additionally, as the risk-free rate was 5 percent, the equilibrium price of the stock was $14 (i.e., $0.70 ÷ 0.05). Therefore, if the trading price was more than $14 per share, the stock would be in a bubble. On average, the median stock price during the trials was $64.30, demonstrating a massive bubble that eventually crashed to a median stock price of $14.13. Using fMRI, these scholars found that the nucleus accumbens tracked the price of the stocks, as it calculated the potential reward of the stock and predicted future price changes and crashes. Additionally, they found that the anterior insular cortex was associated with a higher tendency to sell for people who earned more money, suggesting that it may act as a warning signal to these subjects.

Home ownership provides another set of critical financial decisions. Seiler and Walden used a sample of 20 homeowners who had an existing mortgage in a mid-sized city in the southwest.[9] These homeowners were required to read a list of loan characteristics associated with hypothetical mortgages and answer how likely they might be to strategically default (i.e., they can afford to make the monthly payments but still choose to default) and walk away from their mortgage under different financial conditions. For example, consider that you purchased a home for a selling price of $100,000 using a down payment of $10,000. Now, you still owe $85,000 on the mortgage. Would you strategically default if your home's current price was only $55,000 (a negative $30,000 equity position)? What if under different circumstances you had a negative equity position of $185,000 on a home you purchased for $280,000? The authors found no neurological findings between trials that compare down payment versus negative equity positions, demonstrating that the subjects did not fall victim to the sunk cost fallacy. They also examined a cognitive dissonance theory, that is, even though it may be financially beneficial to default on a mortgage that has a negative equity position, people will still avoid defaulting because it causes uncomfortable psychological conditions due to moral and ethical issues. The results showed that the anterior cingulate cortex was significantly activated when subjects were presented with scenarios in which they were substantially underwater on their mortgage, suggesting that this region of the brain was responsible for moral objections. They also examined trials in which people decided they would default versus scenarios in which they would not default. They found that when the homeowners elected to default, they showed more activation in the lingual gyrus and motor cortex. Other research found that these neural regions are activated when people engage in autobiographical thinking, suggesting that the homeowners in this experiment truly imagined themselves in this challenging position when they elected to default.[10]

Evidence from Brain Damage

Many of the neuroeconomic studies examine neural activity during financial tasks. However, to better understand the role of various parts of the brain in decision making, it is essential to examine the extreme cases, like how decisions are made by people with structural damage to their brain. Thus, if a person has such extreme structural differences, they will likely make different decisions. This can help to illustrate how the brain functions in the financial context.

A classic example is patient known as S.M. S.M. is a female who has had complete bilateral amygdala destruction since childhood due to a rare genetic condition. She has little or no capacity to experience fear. She has been featured in many scholarly studies.[11] In general, she has shown no fear of handling snakes or spiders or watching scary movies. She is very outgoing, friendly, and flirtatious, but fails to recognize negative social cues from others. Even though she has been the victim of many traumatic encounters, like being held up at knifepoint and gunpoint, and nearly killed in a domestic violence incident, she does not exhibit depression, desperation, or other signs of associated behavioral responses. Her lack of fear may inhibit her from detecting potential threats. Unfortunately, we know little about her financial and economic decisions. However, clearly her damaged amygdala inhibits her fear and significantly impacts her overall decision making.

One form of brain damage comes from lesions, which are scarred areas on the brain that interfere with normal activity. Bechara and colleagues examine the difference between groups with damage to the ventromedial prefrontal cortex and damage to the amygdala in the Iowa Gambling Task.[12] In the Iowa Gambling Task, the subjects must select a card from one of four card decks (A, B, C, and D). Each card either wins or loses them money. Decks A and B have a 50 percent probability of winning $100 and a 50 percent probability of losing $250. Decks C and D have a 50 percent probability of winning $50 and a 50 percent probability of losing $50. This task attempts to see if subjects have the neural processes available to realize that decks A and B are disadvantageous and thus avoid selecting from them. The results show that people with damage to either the ventromedial prefrontal cortex or the amygdala were impaired in the gambling task and unable to decode that decks A and B were disadvantageous. In fact, they continued to select from the less optimal and riskier decks.

De Martino and colleagues conducted an experiment on loss aversion in subjects with damage to their amygdala and subjects in a control group.[13] In this experiment, the subjects were given $50 and asked to accept or reject a series of gambles with 50–50 odds. The risky gambles ranged from gains and losses of $20 to $50. The authors' results showed that both groups of subjects retained a normal ability to respond to changes in the expected value of the gamble outcomes. However, they also found that people with damage to their amygdala exhibited less loss aversion because they were less emotional about previous gamble outcomes.

Shiv and colleagues examined loss aversion in a sample of people with lesions on various components of their neural circuitry that are critical for the processing of emotions (amygdala, orbitofrontal cortex, right insular cortex).[14] Also included were people with lesions in regions not associated with emotion, and people with no lesions. In this experiment, the authors provided all the subjects with $20 and then had them participate in 20 rounds of investment decisions where they could risk $1 and have a 50 percent chance to win $2.50 or a 50 percent chance to lose the $1 or to make no bet at all. The authors' results showed that people with brain lesions in regions related to emotion are more likely to continue to take the gamble compared to the other participants who became risk averse after losing. Additionally, the people with brain lesions on components of their neural circuitry critical to processing emotion won more money because they were not loss averse to the gamble. This area of trauma is the same area in which psychopaths have deficits, and the results suggest that psychopaths may make better investors as they are less likely to make emotional decisions.

Biological Foundations of Behavioral Finance

While we have learned much about various psychological biases and cognitive errors that impact investing, we are just now learning the source of those biases. Kuhnen and Knutson developed an experiment using fMRI to find the neural basis of deviations from rational risk taking by examining risk-seeking mistakes (where subjects take risks they should not) and risk aversion mistakes (where subjects do not take risks when they should).[15] In this experiment, the participants had to invest in one asset after seeing two stocks and a bond. Following the selection, one stock would consistently outperform the other stock, while the bond would always gain a risk-free $1. Having seen the outcome of the trial, the participants then selected one of the assets 19 more times. Each decision might result in an optimal (rational) decision or in three types of mistakes. Risk-seeking mistakes are characterized as picking a stock when the bond was the optimal choice. Risk aversion mistakes are defined as picking a stock when selecting the bond was optimal. Finally, confusion mistakes are categorized as selecting a stock when the other stock is the optimal choice. The participants' results showed that the nucleus accumbens, which is related to managing risk and reward, was active during risky choices as well as risk-seeking mistakes. Additionally, the anterior insula, which helps to regulate emotions and avoidance behavior, was active during risk aversion mistakes. This suggests that the subjects wrongly chose to be risk averse owing to an emotional response.

Similarly, Breiter and colleagues conducted an experiment to examine neural responses to gains and losses.[16] In this experiment, the subjects were presented three spinners. The first spinner was "good," and had an equal chance of a gain of $10, $2.50 or $0. The second spinner was "bad" and resulted in an equal chance of a loss of $6, $1.50, or $0. The final spinner was "intermediate" and had an equal chance of a gain of $2.50, a loss of $1.50, or neither gain nor loss. The losses are always less than the gains to be consistent with prospect theory. The subjects then went through 19 trials in which they randomly selected one of the spinners and one of the outcomes for that spinner. When subjects won money or lost money, their nucleus accumbens, amygdala, and hypothalamus were active. This result shows that the subjects had an emotional response to earning or losing money (amygdala and hypothalamus activation), which is related to reward processing (nucleus accumbens activation).

The framing effect is a cognitive bias whereby individuals react differently depending on how a question is asked. For example, De Martino and colleagues conducted an experiment with college students in which they provided the students £50 at the beginning.[17] Students then had to select between a sure option and a gamble which was presented in the context of two different frames. The sure option was written as a "gain" frame whereby they could keep £20 of the £50 or in a "loss" frame whereby they would lose £30 of the £50. A simple analysis of comparing the two options shows that they are exactly the same as keeping £20 and losing £30—both provide the participant with £20. However, due to the framing effect, people are much more likely to select the sure option when it is written in the context of a "gain." Additionally, the "gamble" option was identical in both frames in that they would either keep the £50 with a 40 percent probability or lose all of it with a 60 percent probability. The expected outcomes of the gamble and the sure options were equivalent. The results showed that the amygdala was activated during frame-consistent choices such as risk seeking during losses and risk avoidance in gains. This makes sense because the amygdala is associated with fear and anxiety, suggesting that the participants were making emotional decisions.

Emotions appear to be a significant influence in loss aversion. Sokol-Hessner and colleagues conducted an experiment in which people had to make choices with two different cognitive strategies.[18] Each participant was told to consider each decision in isolation (i.e., this choice is the only decision you will make), called the "attend" treatment, and then make decisions while considering each decision in the greater context (i.e., each choice is one of many or part of a portfolio), called the "regulate" treatment. It was emphasized for the regulate treatment to keep a running tally of previous outcomes and overall earnings. This was done to encourage the participants to regulate their emotions. The choices that the two treatments selected had win/loss outcomes that varied randomly across all trials. The results showed that emotional regulation strategies can modify behavior and physiological responses to the emotional stimuli of gains and losses. Specifically, the regulated treatment showed reduced loss aversion and a decrease in the neural correlates of the responses in the amygdala and striatum regions. Thus, these authors argued that it is possible to learn to make less emotional decisions and be a more rational investor.

Another experiment also investigated the framing effect. However, this time participants started with 100 points and had to decide between option A whereby they kept 80 points or option B whereby they lost 20 points. The findings show that the dorsal anterior cingulate cortex activates during frame-inconsistent choices, such as risk avoidance for losses and risk seeking for gains, as this region of the brain is associated with effortful control and conflict monitoring.[19] Moreover, Li and colleagues conducted an experiment in which they provided people with $40 to start, and the subjects had to select from two choices.[20] The first choice was option A that stated that they could keep $30 guaranteed or participate in a gamble whereby they had a 75 percent chance of keeping it all and a 25 percent chance of losing everything. Option B stated that they could lose $10 guaranteed or participate in a gamble where they had a 75 percent chance of keeping it all and a 25 percent chance of losing everything. The authors' results showed that the irrational choice of frame-inconsistant decisions was due to neural profiles associated with the temporal brain, such as the posterior cingulate cortex rather than with neural profiles associated with emotion (the amygdala). This implies that people are acting irrationally due to their subconscious, not emotion, which is a significant contribution as most research had argued that emotion causes irrational behavior. Furthermore, the same study found that the rational choice of frame-inconsistent decisions was most correlated with effortful processing, such as using reason.

Cary Frydman conducted an experiment to examine the neural activity from the realization of portfolio choice.[21] Specifically, the scholar was interested in how the brain views increases in personal wealth and in the wealth of peers. There is a debate in economics whether people value the level of their own wealth or whether they value the level of their wealth relative to the wealth of others. In this experiment, subjects started with $100 in experimental cash and had the opportunity to invest in the risk-free asset that paid nothing or a risky asset that was based on historical markets over the course of 200 trials. After making their allocation, the subjects saw the realized return of each asset and an updated portfolio value. Finally, the screen revealed the investment allocation of peers' portfolios to examine how other subjects' portfolios influenced the subjects' future allocations. The results showed that neural activity in the ventral striatum increases when a subject's own wealth increases but decreases when a peer's wealth increases. In economic terms, there is a biological reason why people have increased utility when their wealth increases and decreased utility when a peer's wealth increases. Subjects were more sensitive to falling behind their peers than getting ahead themselves. Thus, people have a preference for relative wealth rather than absolute wealth.

The endowment effect is the tendency to place a higher value on items that you own. For example, Carmon and Ariely conducted an experiment in which subjects either had hypothetical NCAA final four tournament tickets and were trying to sell them or did not have tickets and were trying to buy them. The authors found that subjects tried to sell the NCAA final four tournament tickets at a price 14 times higher than the price for which they would be willing to buy the same ticket.[22] This is a typical example of the endowment effect.

Knutson and colleagues developed an experiment to find the neural antecedents of the endowment effect using fMRI.[23] In this experiment, subjects were given $20 in cash to potentially buy products. Two products were randomly assigned to potentially sell. The three products that were used in the experiment were an iPod Shuffle, a 2GB flash drive, and a digital camera. After this initial endowment, the subjects went through "buy," "sell," and "choose" trials. In the "buy" and "sell" trials, the subjects saw one of the two products they could buy or sell for one of 18 possible prices. During the "choose" trial, the subjects saw one of the two products and one of the 18 possible prices and were asked to choose between the item or the associated cash. Because the "choose" and "sell" conditions produced the same outcome, receive money or the product, the difference in their response would elicit the framing effect. The authors found that the nucleus accumbens was activated during both buying and selling and not related to the endowment effect. Furthermore, they found that the prefrontal cortex was activated during buying and that the right insula was activated during selling. This demonstrates that these two neural regions are related to the irrational decisions of the endowment effect.

Another psychological bias is the disposition effect, which is the irrational decision to sell stocks that have appreciated and to keep stocks that have depreciated. Research shows that people who commit the disposition effect earn lower returns as the stocks they sell continue to do well while the stocks they keep continue to do poorly. Frydman and colleagues use fMRI to examine the neural activity of subjects committing the disposition effect.[24] They found that the ventromedial prefrontal cortex and the ventral striatum were active during decisions involving a capital gain as the subjects analyzed the risks and rewards of their decisions. However, they did not find strong evidence of neural regions related to capital loss decisions because there were only three realized losses per participant and the model could not achieve statistical power due to the small sample size.

Regret is an emotion that can make investors act irrationally. For example, in 2002, Apple had a stock price of $14.33 per share. In 2003, the stock price jumped by nearly 50 percent to $21.37 per share, and in 2004 it jumped by more than 200 percent to $64.40 per share. Many investors may have regretted not purchasing Apple stock in 2002, 2003, or 2004 and will not buy Apple stock because of that regret. This is called the regret of omission. In hindsight, not buying Apple stock during this period would have been a big mistake because Apple's stock price rose by 1739 percent from 2005 to 2018, while the S&P 500 increased by just under 100 percent.

Frydman and Camerer used fMRI tools to examine investor regret when they observed a positive return for a stock that they decided not to purchase.[25] In this experiment, people had the opportunity to trade three stocks:

A, B, and C. At the beginning of the experiment, the participant was provided with $350 in experimental currency and required to buy one share of each of the three stocks for $100 each. Every subject would undergo 108 trials where there was a price update followed by a trading decision. Following the price update, the participants were randomly given one of the three stocks and asked if they wanted to trade the stock. The authors then examined neural activity for repurchase decisions and the regret of not trading the stock in previous trials. The results showed a decrease in neural activity in the ventral striatum when the stock that they had recently sold increased in price. This result suggested "fictive learning," which is the brain's response to a reward that could have been had, but which was not chosen, and results in regret. They also found that the ventromedial prefrontal cortex, which is part of the frontal lobe, activates when presented with the option to repurchase stock. Additionally, they found a strong relationship between repurchasing mistakes and the disposition effect, which suggests that there may be a common psychological mechanism between regret and loss aversion. One possible explanation is that investors irrationally believe that prices will revert back to their mean, which means that buying losers and selling recent winners would be a profitable strategy, resulting in the disposition effect and repurchase effect.

Genoeconomics

Genoeconomics Research

Genomic research has become an important part of the general population as many people have purchased kits to test their DNA for information about their ancestors' countries of origin. Additionally, genes determine thousands of traits like eye or hair color and blood type. Furthermore, genes can provide information about health risks, such as celiac disease (HLA-DQB1 and HLA-DQA1 genes) or Parkinson's disease (LRRK2 and GBA genes). However, it may come as a surprise that certain genes predispose a person toward certain financial behavior.

Genes and Receptors

Receptors are proteins in a cell membrane that a neurotransmitter or hormone must bind with in order to trigger a response in that cell. Genes of the chromosomes code for all proteins including receptors. Alleles are different variations of the gene. Consider eyes: the trait is eye color, but different alleles are different variations of the gene, which will result in different eye colors like blue or brown.

The dopamine receptor D4 (DRD4) is one of the most common dopamine receptors, which may repeat the sequence of nucleotides of the DNA between two and 11 times. About three-quarters of the population carries the 4-repeat or the 7-repeat sequences of the allele. There is also a wide variation in the distribution of allele repeat sequences within the population of different cultures. The variations in the DRD4 have been linked to addictive behaviors, ADHD, bipolar disorder, eating disorders, and other psychiatric conditions. Different alleles with longer repeats (seven or more) are associated with reduced sensitivity to dopamine. A person with an allele for a larger number of repeats would need higher levels of dopamine stimulation to induce the same internal reward compared to someone with a shorter number of repeats. The higher number of repeated sequences of DNA means that there are more amino acids in the protein receptor causing it to fold differently in the cell membrane and establishing a slightly different shape, which lowers the affinity of its binding site (dopamine does not bind as strongly) causing reduced sensitivity. Thus, it is important to study the role of DRD4 in economic decision making.

The 5-HTTLPR is the gene that codes for a serotonin transporter to move serotonin into a cell. The gene has two variations: a short or long allele. Research suggests that the long allele results in higher serotonin levels and has its greatest impact on the amygdala. This suggests that lower serotonin levels may cause risk-averse behavior.

Monoamine oxidase (MAOA) is an enzyme that regulates the breakdown of monoamines, such as serotonin, dopamine, norepinephrine, and epinephrine (aka adrenaline which triggers the body's fight or flight response). Carriers of the 3.5 or 4 repeats (MOA-H) gene allele exhibit higher expression of the enzyme compared to carriers of the 2, 3, or 5 repeats (MAOA-L) allele. Therefore, people with these different alleles may break down dopamine and serotonin differently, and thus make different financial decisions.

DRD4 and Financial Decision Making

Carpenter and colleagues developed an experiment with college students whereby each student faced three risky decisions in the same order.[26] For each choice, the students were presented with a ring of six possible gambles. The first choice was to select a 50–50 gamble between positive outcomes. Would you rather pick (1) a 50–50 chance of winning $92 or $0; (2) a 50–50 chance of winning $62 or $18; or (3) a 50–50 chance of winning $33 or $33? The second choice was a loss choice. For example, would you select a 50–50 chance of (1) winning $41 or losing $46; (2) winning $12 or losing $32; or (3) losing $25 or $3? The final selection was ambiguous. The students selected between gambles of identical payouts as the first choice; however, the probability of outcomes was unknown (i.e., ambiguous). The results showed that individuals with the DRD4 7R+ gene were more likely to select the riskier choices in all three scenarios, suggesting that people with reduced sensitivity to dopamine will take more risks.

Dreber and colleagues examined the impact of the DRD4 gene and risk taking in men. In this experiment, the authors had their participants select a percentage of their endowment in a 50–50 coin flip.[27] If they guessed the side of the coin that was facing up, they would earn 2.5 times their gamble, and if they guessed wrong they would lose their gamble. Thus, this gamble had a positive expected return as they would either profit by 250 percent or lose 100 percent of the gamble. Consistent with the notion that reduced sensitivity to dopamine leads to greater risk taking, the authors found that individuals who carried the DRD4 7R+ gene were willing to wager about 30 percent more on the gamble.

Kuhnen and Chiao examined the impact of the DRD4 and 5-HTTLPR alleles on financial risk taking.[28] The subjects participated in an investment task whereby they were given the choice of betting a portion of their endowment (a sum between $23 and $28) on a risky option such as a 50–50 chance of winning 23 percent of their gamble or losing 13 percent of their gamble or a riskless option of a guaranteed profit of 3 percent. The authors found that people who had the long allele of the 5-HTTLPR chose to gamble about $2.69 more than those with the short allele. Additionally, the authors found that those with the DRD4 7R+ gene were willing to wager about $2.46 more than those with other DRD4 genes.

MAOA and Financial Decision Making

Frydman and colleagues conducted an experiment whereby subjects were endowed with $25 and had to complete 140 decision trials.[29] In each trial, they had either a $0 gain with 100 percent certainty or a gamble that had a 50 percent chance of winning $7 or losing $4. They had no feedback from previous outcomes. Their results showed that people who carry the MAOA-L gene accepted the gamble 5 percent more frequently than those with the MAOA-H allele, while finding no difference in the gambling propensity of dopamine and serotonin genes. Accordingly, their results showed that people who have the MAOA-L gene are more likely to exhibit better financial decisions under risk because they are less impulsive.

Zhong and colleagues examined the impact of the MAOA gene on risk attitudes. In their experiment, the subjects took part in a lottery-style task whereby they could either have a 1 percent chance of winning $200, a 10 percent chance of winning $20, or a 100 percent chance of winning $2.[30] Additionally, they conducted a second lottery-style task whereby the subjects would either lose $2 for sure or have a 0.1 percent chance of losing $200. The authors found that subjects who had the MAOA-H allele had a preference for the long-shot lottery and selected the chances of winning or losing $200 more than carriers of the MAOA-L allele.

De Neve and Fowler examined the impact of the MAOA gene on financial behavior.[31] The authors used the National Longitudinal Study of Adolescent Health that surveys people from adolescence through their young adult years. In their early adult lives, the participants were genotyped, which included looking at the MAOA gene, and asked about their credit card debt.

The authors merged this data with demographic information to examine if the MAOA gene impacted people's credit card borrowing behavior. They found that having the MAOA-L gene was linked to increasing the likelihood of having credit card debt by about 4 percent.

Summary

Neuroeconomics analyzes the structure and function of the brain during financial and economic decisions to help to explain human behavior. The research in this chapter provides evidence against traditional finance's assumption that all individuals make rational decisions by mapping where irrational choice occurs in the brain. For example, decisions related to the frontal lobe (controlled processes) show more rational choice, such as being able to disentangle the advantageous and disadvantageous decks in the Iowa Gambling Task and during stock buying and selection decisions. Conversely, the amygdala activates during emotional and fear responses and is an automatic process. Research shows that individuals who use their amygdala during financial decisions are more likely to make irrational decisions, such as avoiding ambiguous choices, being impaired during the Iowa Gambling Task, and making poor decisions due to loss aversion. Furthermore, the nucleus accumbens is part of the dopaminergic pathway that mediates the risk and reward of decisions. It is active during stock buying and selling and after losses and gains. Finally, genetics appears to have an influence on how the brain processes dopamine and serotonin with neurotransmitter receptors that increase the amount of dopamine and serotonin in the brain causing increased risk-taking behavior.

Questions

1   What are the roles of the frontal lobe, the amygdala, the nucleus accumbens, dopamine, and serotonin?
2   What is the difference between brain automatic and controlled processes?
3   How do the frontal lobe, amygdala, and nucleus accumbens relate to financial decisions?
4   How are the dopamine receptor, serotonin transporter, and monoamine oxidase enzyme related to financial decisions?

Notes

1   Richard M. Shiffrin and Walter Schneider. 1977. Controlled and automatic human information processing: II. Perceptual learning, automatic attending and a general theory. *Psychological Review* 84:2, 127.
2   Daeyeol Lee, Hyojung Seo and Min Whan Jung. 2012. Neural basis of reinforcement learning and decision making. *Annual Review of Neuroscience* 35, 287–308.
3   César Valentim de Oliveira Carvalho Júnior, Edgard Cornacchione, Armando Freitas da Rocha, and Fábio Theoto Rocha. 2017. Cognitive brain mapping of

auditors and accountants in going concern judgments. *Revista Contabilidade & Finanças* 28:73, 132–147.

4 Hengyi Rao, Marc Korczykowski, John Pluta, Angela Hoang, and John A. Detre. 2008. Neural correlates of voluntary and involuntary risk taking in the human brain: An fMRI Study of the Balloon Analog Risk Task (BART). *Neuroimage* 42:2, 902–910.

5 Scott A. Huettel, C. Jill Stowe, Evan M. Gordon, Brent T. Warner, and Michael L. Platt. 2006. Neural signatures of economic preferences for risk and ambiguity. *Neuron* 49:5, 765–775.

6 Ming Hsu, Meghana Bhatt, Ralph Adolphs, Daniel Tranel, and Colin F. Camerer. 2005. Neural systems responding to degrees of uncertainty in human decision-making. *Science* 310:5754, 1680–1683.

7 Constantinos Antoniou, Richard D.F. Harris, and Ruogu Zhang. 2015. Ambiguity aversion and stock market participation: An empirical analysis. *Journal of Banking & Finance* 58, 57–70.

8 Alec Smith, Terry Lohrenz, Justin King, P. Read Montague, and Colin F. Camerer. 2014. Irrational exuberance and neural crash warning signals during endogenous experimental market bubbles. *Proceedings of the National Academy of Sciences* 111:29, 10503–10508.

9 Michael J. Seiler and Eric Walden. 2015. A neurological explanation of strategic mortgage default. *The Journal of Real Estate Finance and Economics* 51:2, 215–230.

10 Russell N. James III and Michael W. O'Boyle. 2014. Charitable estate planning as visualized autobiography: An fMRI study of its neural correlates. *Nonprofit and Voluntary Sector Quarterly* 43:2, 355–373.

11 Ralph Adolphs, Daniel Tranel, Hanna Damasio, and Antonio Damasio. 1994. Impaired recognition of emotion in facial expressions following bilateral damage to the human amygdala. *Nature* 372:6507, 669–72; Adam K. Anderson and Elizabeth A. Phelps. 2001. Lesions of the human amygdala impair enhanced perception of emotionally salient events. *Nature* 411:6835, 305–309.

12 Antoine Bechara, Hanna Damasio, Antonio R. Damasio, and Gregory P. Lee. 1999. Different contributions of the human amygdala and ventromedial prefrontal cortex to decision-making. *Journal of Neuroscience* 19:13, 5473–5481.

13 Benedetto De Martino, Colin F. Camerer, and Ralph Adolphs. 2010. Amygdala damage eliminates monetary loss aversion. *Proceedings of the National Academy of Sciences* 107:8, 3788–3792.

14 Baba Shiv, George Loewenstein, Antoine Bechara, Hanna Damasio, and Antonio R. Damasio. 2005. Investment behavior and the negative side of emotion. *Psychological Science* 16:6, 435–439.

15 Camelia M. Kuhnen and Brian Knutson. 2005. The neural basis of financial risk taking. *Neuron* 47:5, 763–770.

16 Hans C. Breiter, Itzhak Aharon, Daniel Kahneman, Anders Dale and Peter Shizgal. 2001. Functional imaging of neural responses to expectancy and experience of monetary gains and losses. *Neuron* 30:2, 619–639.

17 Benedetto De Martino, Dharshan Kumaran, Ben Seymour, and Raymond J. Dolan. 2006. Frames, biases, and rational decision-making in the human brain. *Science* 313:5787, 684–687; Jonathan P. Roiser, Benedetto de Martino, Geoffrey C.Y. Tan, Dharshan Kumaran, Ben Seymour, Nicholas W. Wood, and Raymond J. Dolan. 2009. A genetically mediated bias in decision making driven by failure of amygdala control. *Journal of Neuroscience* 29:18, 5985–5991.

18 Peter Sokol-Hessner, Colin F. Camerer, and Elizabeth A. Phelps. 2013. Emotion regulation reduces loss aversion and decreases amygdala responses to losses. *Social Cognitive Affective Neuroscience* 8:3, 341–350.

19 Pengfei Xu, Ruolei Gu, Lucas S. Broster, Runguo Wu, Nicholas T. Van Dam, Yang Jiang, Jin Fan and Yue-jia Luo. 2013. Neural basis of emotional decision making in trait anxiety. *Journal of Neuroscience* 33:47, 18641–18653.
20 Rosa Li, David V. Smith, John A. Clithero, Vinod Venkatraman, R. McKell Carter, and Scott A. Huettel. 2017. Reason's enemy is not emotion: Engagement of cognitive control networks explains biases in gain/loss framing. *Journal of Neuroscience* 37:13, 3588–3598.
21 Cary Frydman. 2016. Relative wealth concerns in portfolio choice: Neural and behavioral evidence. Working Paper, University of Southern California, Marshall School of Business.
22 Ziv Carmon and Dan Ariely. 2000. Focusing on the forgone: How value can appear so different to buyers and sellers. *Journal of Consumer Research* 27:3, 360–370.
23 Brain Knutson, G. Elliott Wimmer, Scott Rick, Nick G. Hollon, Drazen Prelec, and George Loewenstein. 2008. Neural antecedents of the endowment effect. *Neuron* 58:5, 814–822.
24 Cary Frydman, Nicholas Barberis, Colin Camerer, Peter Bossaerts, and Antonio Rangel. 2014. Using neural data to test a theory of investor behavior: An application to realization utility. *Journal of Finance* 69:2, 907–946.
25 Cary Frydman and Colin Camerer. 2016. Neural evidence of regret and its implications for investor behavior. *The Review of Financial Studies* 29:11, 3108–3139.
26 Jeffrey P. Carpenter, Justin R. Garcia, and J. Koji Lum. 2011. Dopamine receptor genes predict risk preferences, time preferences, and related economic choices. *Journal of Risk and Uncertainty* 42:3, 233–261.
27 Anna Dreber, Coren L. Apicella, Dan T.A. Eisenberg, Justin R. Garcia, Richard S. Zamore, J. Koji Lum, and Benjamin Campbell. 2009. The 7R polymorphism in the dopamine receptor D4 gene (DRD4) is associated with financial risk taking in men. *Evolution and Human Behavior* 30:2, 85–92.
28 Camelia M. Kuhnen and Joan Y. Chiao. 2009. Genetic determinants of financial risk taking. *PloS One* 4:2, e4362.
29 Cary Frydman, Colin Camerer, Peter Bossaerts, and Antonio Rangel. 2010. MAOA-L carriers are better at making optimal financial decisions under risk. *Proceedings of the Royal Society B: Biological Sciences* 278:1714, 2053–2059.
30 Songfa Zhong, Salomon Israel, Hong Xue, Richard P. Ebstein, and Soo Hong Chew. 2009. Monoamine oxidase A gene (MAOA) associated with attitude towards longshot risks. *PLoS One* 4:12, e8516.
31 Jan-Emmanuel De Neve and James H. Fowler. 2014. Credit card borrowing and the monoamine oxidase A (MAOA) gene. *Journal of Economic Behavior & Organization* 107:B, 428–439.

6 The Influence of Hormones on Financial Risk Taking

Hormones affect moods, emotions, and impulses, which in turn impact behavior. It is well known that hormones like testosterone can influence teenagers to make poor and suboptimal social choices. Yet hormone levels impact peoples' decisions throughout their lives. It should not be a surprise to learn that they also influence financial decisions. This chapter explains how hormonal activity can shift risk preferences and cognitive function, and how these changes influence financial decisions.

Hormones

Hormonal Physiology

Hormones are chemical messengers created by endocrine glands all over the body that control, regulate, and coordinate bodily activities, including the way the neurons of the brain send and receive signals. There are about 50 different hormones distributed throughout the body through the blood and lymph system. These hormones are designed to affect processes in the body, like metabolism, mood, sexual function, growth and development, etc. Hormone imbalances can cause significant problems, like weight gain, skin changes, mood changes, and low energy. While there are several kinds of hormones, this chapter will focus solely on testosterone (the main male hormone) and cortisol (the stress hormone). The finance and business profession is highly male-dominated and given the gender differences in testosterone production, testosterone is a highly researched hormone.

Testosterone is produced in the testes by interstitial cells (also known as Leydig cells) in males starting at puberty and continuing for the rest of their lives. Testosterone production tends to slowly decrease in men after about 30 years. In both men and women, the adrenal cortex also produces small amounts of testosterone continuously from birth to death. Thus, both men and women have testosterone, although the level is much higher in men. In both genders, the hypothalamus produces the corticotrophin releasing hormone (CRH) that triggers the anterior pituitary gland to produce the adrenocorticotropic hormone (ACTH), which triggers the adrenal cortex to produce small amounts of testosterone and estrogen.

Cortisol is known as the stress hormone because it assists the body in responding to stress by helping to maintain normal glucose levels. When a person is stressed, the amount of cortisol present in the body increases, which leads to a rise in glucose levels in the blood. Under normal conditions, cortisol and the adrenal cortex-produced testosterone (not testicular-produced testosterone) have the same hormonal tree, which means that the hypothalamus produces CRH that triggers the anterior pituitary gland to produce ACTH to produce cortisol and testosterone in the adrenal cortex.

When a person is stressed, cortisol has a different function. Whether a person has either short-term stress such as a car accident, is scared, anxious, upset, etc., or has long-term stress such as a chronically sick family member or struggling to pay bills, the nervous system sends a signal to the adrenal cortex to dramatically increase the level of cortisol well above normal. This results in an elevation of glucose levels in the blood. In stressful situations this increase helps to provide cells with the glucose necessary for energy production. However, a long-term increase in cortisol can have serious side effects because it causes protein (such as bone or muscle protein) to break down and be converted to glucose. This can lead to medical disorders such as osteoporosis or muscular atrophy. Additionally, cortisol (including that in cortisone pills or cream) inhibits the immune system leading to a higher susceptibility to infections. Therefore, the more stressed individuals are, the more likely they are to be sick or to have more serious medical conditions. Cortisol levels will rise when the amount of glucose in the blood is low, such as when one is hungry or exercising. The low glucose level triggers an increase in the release of cortisol from the adrenal glands which in turn trigger the release of more glucose until levels return to normal.

Finance professionals, such as traders, executives, and fund managers, are often under an excessive amount of stress. Thus, understanding the impact of cortisol on financial decisions is important to people, firms, and the markets. Oberlechner and Nimgade used a sample of 326 foreign exchange traders and asked them what their greatest sources of stress were. Their results show that "pressure to achieve the profit goal, long working hours, and time pressure" are consistently large sources of stress.[1]

After secretion, steroid hormones such as testosterone and cortisol enter the bloodstream and cross the blood-brain barrier. These hormones alter the production and function of synapses in neurons and the structure of brain cells. This implies that hormones can either stimulate or inhibit neural activity. Cortisol and testosterone receptors are located throughout the brain, including the frontal cortex, amygdala, hippocampus, and nucleus accumbens. For example, testosterone and cortisol can affect the release of the neurotransmitter dopamine, which in turn influences the brain's ability to manage situational risk and return. Therefore, as most would suspect, hormone levels can alter peoples' decision making in general, including financial decisions.

Circulating Testosterone and Cortisol Levels

Testosterone and Economic Risk Taking

One of the most important financial questions in the topical area of physiological finance is how hormones can impact monetary risk taking. However, due to the high cost of collecting and processing hormone samples and paying participants, the number of experiment studies is low.

Apicella and colleagues examine this question using a male-only sample to analyze how circulating levels of testosterone affect people in an investment game originally used by Gneezy and Potters.[2] In this experiment, each person started with $250 and was asked to select a percentage of their holdings to allocate to the risky asset and keep the rest. In order to determine the outcome of the gamble, they flipped a coin. If they successfully guessed which side the coin would land on, they won 2.5 times their gamble, but if they guessed incorrectly, they lost their gamble. Note that there is a substantial risk premium to take part in the gamble. If they invested all $250, their expected outcome would be a 50–50 chance of $625 (= $250 × 2.5) or 0 resulting in an expected return of $312.5 (= $625 × 0.5 + $0 × 0.5), which is much higher than the original $250. Therefore, the amount each person chose to risk can be a measure of risk taking. Overall, the study showed that, on average, people who had higher levels of circulating testosterone levels invested more money in the risky gamble and earned more money.[3]

Stanton and colleagues examined the impact of testosterone on risk taking by employing the Iowa Gambling Task in a sample of men and women.[4] In the Iowa Gambling Task, the subjects must select a card from one of four card decks (A, B, C, and D). Each card either wins or loses them money. Decks A and B have a 50 percent probability of winning $100 and a 50 percent probability of losing $250. Decks C and D have a 50 percent probability of winning $50 and a 50 percent probability of losing $50. This task seeks to discover if subjects have the neural processes available to realize that decks A and B are disadvantageous and thus avoid selecting from them. Their results show that higher testosterone levels are associated with cards selected from the more risky and less advantageous decks. This result shows that while testosterone does increase risk taking, it may do so even if it is likely to result in a detrimental outcome.

Most of the previous studies assume that higher testosterone levels result in higher risk. However, Stanton and colleagues examined the impact of low and high testosterone levels in 298 men and women using risk preference and ambiguity preference tasks.[5] For example, in the risk preference task, people selected between two gambles, such as a 50 percent chance of winning $13 and a 50 percent chance of $0, or a 100 percent chance of $5. Similarly, the ambiguity preference task phrased questions such as having a 100 percent chance of winning $5 or an unknown probability of winning either $13 or $0. Their results reveal a U-shaped pattern. Individuals with either low or high levels of testosterone are more likely to be risk and ambiguity neutral, while those with intermediate levels of testosterone are more likely to be risk and ambiguity averse with similar results for both men and women.

In addition to examining how testosterone affects monetary risk taking, Greene and colleagues examined the impact that testosterone has on career choice.[6] These authors surveyed 1,199 adult males in Australia and found that people who had higher levels of testosterone were more likely to be self-employed suggesting that those with higher testosterone levels may self-select into more risky careers.

Cortisol and Economic Risk Taking

In contrast to testosterone, it is intuitive that cortisol would be inversely related to risk taking. As cortisol levels are higher during times of stress, it is unlikely that a person would want to take on more risk to add stress to their life. Taking new risks may increase stress even more, and thus cortisol may act to impact those decisions by inhibiting risk appetite.

Van Honk and colleagues employed the Iowa Gambling Task to examine the impact of cortisol levels on risk taking.[7] The authors found that people with low levels of cortisol were more likely to take a risk and select from the advantageous decks. These results suggest that higher levels of cortisol are related to risk aversion and may balance people's sensitivity for punishment and reward.

In another study, Kandasamy and colleagues conducted an experiment whereby subjects took part in a lottery-type task.[8] One group was administered cortisol and the other group received a placebo. One of the lottery choices was whether the participant would rather take gamble A that has a 90 percent chance of winning $30 and a 10 percent chance of winning $90, or gamble B that has a 40 percent chance of winning $90, a 40 percent chance of winning $30, and a 20 percent chance of winning nothing. Overall, the authors found that those who were administered cortisol before the experiment exhibited greater risk aversion and would rather select gamble A. This result is consistent with medical literature that shows that when people are more stressed (and have higher cortisol levels), they are less likely to take risk.

The Dual Hormone Hypothesis

If testosterone is related to increased risk taking and cortisol is related to risk aversion, how do these two hormones work together? The dual hormone hypothesis proposes that cortisol mediates the effects of testosterone on behavior. For example, it asserts that high testosterone levels will result in risky behavior only if cortisol levels are low because cortisol suppresses cortical and subcortical neural communications that control socially aggressive behavior. Higher testosterone levels increase risk taking but that influence on risk is mitigated with higher cortisol levels, because high cortisol reduces risk taking, thus creating a better balance. Research has shown support for this hypothesis in social settings as the imbalance in the ratio of testosterone-to-cortisol is associated with social aggression,[9] social dominance,[10] anger,[11] and social risk attitudes.[12]

As it relates to economic risk taking, Mehta and colleagues used the Balloon Analog Risk Task (BART) to study the dual hormone hypothesis in an economic setting.[13] The BART is an economic test in which the subject earns money every time he clicks a button to inflate a balloon. There is an unknown threshold whereby if the subject chooses to inflate the balloon one more time it will pop and he will lose all the money he earned in that round. Therefore, every click to inflate the balloon increases the amount of money the subject can earn, but simultaneously increases the risk that the balloon will pop. Suppose you have already clicked the inflate button several times and have earned $45. Do you click it again to get to the next level, risking it all? The researchers' results found consistent evidence of higher risk taking in people with high testosterone levels when cortisol levels are low, but no evidence that testosterone affects risk taking when cortisol levels are high, which is consistent with the dual hormone hypothesis.

Testosterone, Cortisol, and Financial Decisions

As with economic risk taking, there is limited evidence on how circulating hormone levels affect financial decisions. Coates and Herbert are among the first to examine the impact that testosterone has on financial trading.[14] These scholars used a sample of 17 male floor traders at the London Stock Exchange. The researchers measured the traders' morning (11:00 a.m.) and afternoon (4:00 p.m.) levels of testosterone for eight consecutive business days while the traders worked under the real conditions of their jobs. The

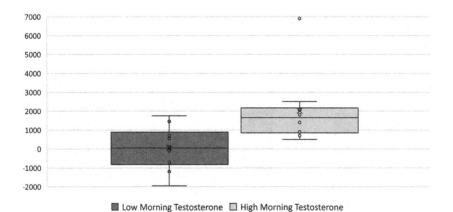

Figure 6.1 Testosterone and Traders' Profits and Losses

researchers found that on days when the traders' morning testosterone levels rose above their own median levels for the eight-day period they also experienced higher profitability, thus suggesting that morning hormonal levels can predict performance. Figure 6.1 illustrates the effect of morning testosterone levels on traders' profits. The figure shows that high levels of testosterone in the morning result in average daily profits of about £1,600 with a range of about £500 to £7,000. Low levels of testosterone in the morning resulted in a profit range of –£1,250 to £1,500 with a lower average of just over £0. Note that the best performance on a low testosterone day was lower than the average performance on a high testosterone day.

Coates and Herbert also examined the association between cortisol and financial trading using the same 17 male floor traders. They found that on days when the market was more volatile and their trading profits and losses had a higher variance, the traders experienced higher cortisol levels. This shows that market volatility stresses the traders. Figure 6.2 illustrates the traders' cortisol levels by the amount of profit and loss (P&L) standard deviation. The figure shows that when market volatility was low, with a P&L standard deviation of under 2.8, investor cortisol (stress) levels also remained low (under 1,625 pg/ml). However, when market volatility was high, with trader P&L standard deviation over 3.6, investor cortisol levels skyrocketed to over 2,000 pg/ml. This demonstrates that market volatility causes stress, even for professional traders.

Nofsinger and colleagues examined the impact of testosterone and cortisol on trading behavior using a financial trading simulation application known as the Rotman Interactive Trader with a sample of students in a masters' finance program.[15] The students were endowed with $500,000 to invest in a portfolio of assets with the goal of having $1.5 million at the end of the simulated 20-year horizon (requiring an average annual return of 5.65 percent). They were able to select from five different electronic trading funds

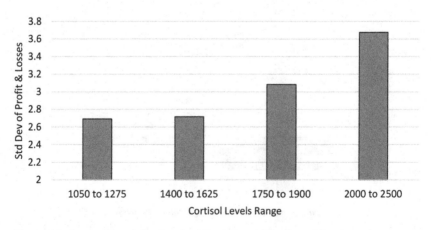

Figure 6.2 Cortisol Levels by Standard Deviation of Profit and Loss

(ETFs) that historically had different returns, volatilities, and correlations with each other. After making their initial portfolio selections, the Rotman Interactive Trader simulated the first five years and displayed the realized portfolio return as well as the realized return for each ETF. Each ETF return was simulated using a random walk with a positive drift. The participants could then reallocate their portfolio for the next five-year period. This continued until the full 20-year period was concluded. Testosterone and cortisol levels were measured both before and after the trading simulation. Figure 6.3 provides a graphical representation of these results.

The results can be summarized as follows:

- People with higher testosterone levels consistently had higher excess returns (expected returns greater than that needed to meet the 20-year goal), suggesting that higher testosterone was associated with higher investment risk taking.
- Higher cortisol levels had lower excess returns, consistent with lower risk taking.
- Subjects with higher portfolio returns had higher changes in their testosterone levels by the end of the experiment. Therefore, as investors made more money, their testosterone levels may have risen, which manifests into more risk taking.

Nofsinger and colleagues also used the same financial simulation software to examine the impact of testosterone and cortisol on investment biases.[16] Specifically, they examined the disposition effect and excessive trading. The disposition effect is the behavior of investors selling their winning positions too soon and holding their losers too long. This is driven by the positive feelings of pride when a winner is sold and avoiding the negative feelings of regret when a loser is sold. This is irrational for three reasons. First, the winner sold continues to perform well, so it was sold too soon. Second, the

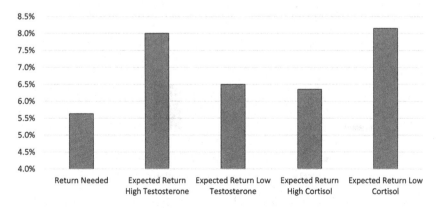

Figure 6.3 Testosterone, Cortisol, and Excess Return

loser being held continues to underperform, so it should have been sold. Finally, capital gains tax means taxes are paid for selling a winner at a capital gain, but a reduction in taxes is likely when a loser is sold for a capital loss. The disposition effect is measured by the proportion of winners sold compared to losers sold. Excessive trading is often used as a proxy for overconfidence and burdens the investor with trading costs. Trading level is measured as portfolio turnover, which is measured as the percentage of the portfolio that is sold for new assets (turned over) during the period. In their experiment, the authors found that both cortisol and testosterone are positively related to irrational financial decisions. Figure 6.4 provides a graphical representation of these results. When looking at the disposition effect, investors with low cortisol levels realized 6 percent more gains than losses, while investors with high cortisol levels realized 27 percent more gains than losses. Note that higher levels of testosterone also resulted in higher levels of disposition effect behavior. Similarly, investors with low cortisol or testosterone levels turned over approximately 14 percent of their portfolio in each time period, while individuals with high testosterone levels turned over their portfolio 22 percent of the time and investors with high cortisol levels turned their portfolio over 27 percent of the time. These results show that high levels of cortisol and testosterone are associated with people who trade irrationally and excessively.

Cueva and colleagues examined the effects of testosterone and cortisol on risk taking and market instability in a sample of 142 people using 15 periods of a double auction trading game.[17] The double auction market means that one investor must submit an "ask" for how much they are willing to pay for the financial asset and another participant must submit a "bid" for how much they are willing to sell it. If a buyer and seller have the same ask and bid then the trade will be executed. If the buyer and seller are not matched

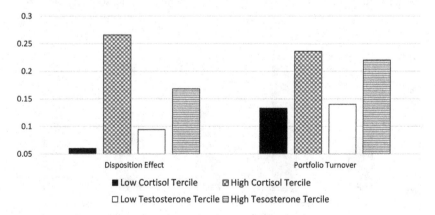

Figure 6.4 Cortisol, Testosterone, and Investment Biases in a Financial Trading Simulation – Realized Gains over Losses (%)

then there will be no trade. The stock they were trading could pay a dividend of $36, $24, $16, or $4 with an equal chance of each plus a payment of $360 at the end of the 15th round. The study's first finding was that circulating cortisol levels, as measured through a saliva test, were associated with greater trading activity in men that created mispriced stock prices and market instability. Thus, people under stress, as measured by their cortisol levels, are more likely to engage in irrational trading activity.

In two other experiments, Cueva and colleagues administered hormones to participants prior to their participation in trading exercises in order to examine the exogenous relationship between the hormones and investment behavior. In one experiment, subjects were given either a cortisol gel or a placebo. In the second experiment, subjects were given either a testosterone gel or a placebo. In these experiments, subjects were given £10 to invest. In a series of 80 trials, they were shown plots of two stocks in each round and were asked to decide how much to invest in each stock. The plots graphically illustrated the stocks' price volatility and trend. A graphic representation of the results can be seen in Figure 6.5. This figure shows that there was little to no difference in selecting low variance stocks when comparing cortisol and testosterone levels to the placebo. However, individuals who were administered cortisol or testosterone were much more likely to purchase highly volatile stocks compared to those administered the placebo. These two experiments suggest that administered cortisol and testosterone lead to higher risk taking.

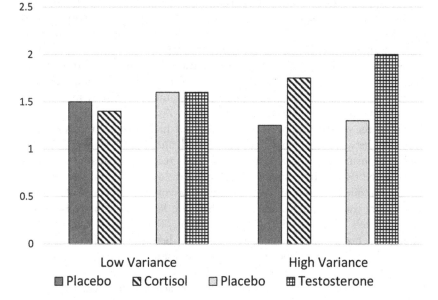

Figure 6.5 Cortisol and Testosterone and Risk Taking in a Double Auction Trading Game – Amount Invested in Low or High Variance Stock, £

Nadler and colleagues examined the impact of testosterone on asset pricing bubbles in a sample in which half the subjects were given a testosterone gel and the other half were given a placebo.[18] This experiment conducted 17 sessions of double auction markets that had three rounds of trading periods each. The asset auctioned paid a dividend of either $18 or $0 with a 50–50 chance, which leads to the expected value of the shares being $9. The authors found that testosterone generated larger and longer lasting bubbles (periods during which the trading price deviated from the expected value). Given that trading is a male-dominated industry, one could speculate that this gender asymmetry may influence the frequency and size of market bubbles, like the dotcom bubble, bitcoin, and others.

Little empirical research has been conducted on the impact of gender mix on trading and market dynamics. However, Bose and colleagues used a mathematical model to theoretically examine the impact of market volatility when the percentage of female traders increases.[19] It assumes that the amount of testosterone among all the traders will decrease. The model suggests that volatility will not decrease if more female traders are present, but volatility may reduce the occurrences of market crashes, which is consistent with the research carried out by Nadler and colleagues (see above).

Testosterone Proxies

2D:4D Ratio and Facial Masculinity

Due to the high costs associated with experiments using circulating hormonal levels and low sample sizes, researchers have examined the question of how hormones influence economic and financial behavior via proxies. Human biology scholars have found that exposure to higher levels of testosterone in the womb has lasting and identifiable physical affects. If testosterone has an impact on some physical characteristics, then it could also impact the formation of the brain. Thus, high levels of prenatal testosterone may impact decision making and risk taking in adults.

One of the most commonly used proxies for high prenatal testosterone is the ratio between the length of the second and fourth fingers (2D:4D). Higher exposure of prenatal testosterone levels results in a smaller difference between the two digits compared to those with lower prenatal testosterone. Another frequently used measure is facial masculinity, which is also based on prenatal testosterone levels. Researchers use facial masculinity by examining people's different width-to-height ratios. This ratio is calculated by taking the bizygomatic width (the distance between the left and right temple) divided by the upper-face height (the distance between the upper lip and midpoint of the inner ends of the eyebrows). A higher ratio (i.e., a larger bizygomatic width and shorter upper-face height) signifies a more masculine face and more exposure to prenatal testosterone. The benefits of using these proxies are that they do not involve the high cost of running hormone tests and thus can utilize much larger sample sizes. Additionally, using facial masculinity, researchers need only a photograph of subjects and tests do not need to be done in an experimental lab.

Testosterone Proxies and Economic Risk Taking

Garbarino and colleagues used the 2D:4D ratio to assess risk taking in a sample of 152 men and women by selecting three lottery choices.[20] For each choice, the person was given the possible gambles of $22 or $22, $30 or $18, $38 or $14, $46 or $10, $54 or $5, and $60 or $0. In the first trial, there was a 50 percent chance of the first outcome happening and a 50 percent chance of the second happening. In the second trial, there was a 75 percent chance of the first payout and a 25 percent chance of the second payout. Finally, in the third trial, there was a 25 percent chance of the first outcome happening and a 75 percent chance of the second happening. Would you make the same selection in all three trials or would your choice change when the probabilities changed? The researchers found that people with a lower 2D:4D ratio (more prenatal testosterone) took more risk, and that this measure of prenatal testosterone could partially explain the variation in financial risk taking between the genders.

As the research shows that testosterone is consistently related to increased risk taking and aggressiveness, Nye and colleagues examined whether higher prenatal testosterone exposure was a determinant in wage differences.[21] The researchers used the Russian Longitudinal Monitory Survey and people's 2D:4D ratio to answer questions with a sample size of over 4,000 people. They found that lower digit ratios (higher testosterone exposure) resulted in higher wages for both men and women after controlling for variables such as age, education, and occupation.

Testosterone Proxies and Financial Decision Making

In addition to Coates' previous research on circulating testosterone and cortisol levels in traders, Coates and colleagues examined the 2D:4D ratio among a sample of 44 high frequency financial traders at the London Stock Exchange.[22] They found that the 2D:4D ratio predicted the traders' long-term profits. Those exposed to higher levels of prenatal testosterone consistently earned higher profits, likely due to more risk taking. Furthermore, they found that higher exposure to prenatal testosterone resulted in career longevity, suggesting that employees in the financial markets may self-select into the field due to their underlying biological traits.

Lu and Teo examined the effect of facial masculinity and hedge fund manager performance.[23] They examined nearly 20,000 different hedge fund managers' performance from 1994 to 2015. This work focused on risk-adjusted returns. That means that while taking more risk may result in higher returns during certain time periods, most investment managers were judged on their risk-adjusted return (i.e., alpha). The scholars found that those fund managers who had higher prenatal testosterone exposure consistently underperformed compared to low testosterone hedge fund managers. Figure 6.6 illustrates the portfolio alpha (risk-adjusted return) depending on the level of hedge fund managers' prenatal testosterone exposure. The figure shows that hedge fund managers who had the least testosterone exposure had an alpha value of nearly 5 percent higher than hedge fund managers who had the highest testosterone exposure. They explained that this poor performance stems from a greater preference for lottery stocks and committing the disposition effect (not selling poor performing stocks). Additionally, they found that managers with high levels of testosterone were also more likely to terminate their funds, had more disclosure violations, and took more operational risk.

Cronqvist and colleagues used the Swedish Twin Registry and obtained information from 70,000 twins about their investment activity.[24] They examined the twin testosterone transfer hypothesis which states that females who share the womb with a male twin are exposed to a larger amount of prenatal testosterone than females who share a womb with a female twin. Therefore, to examine the impact of prenatal testosterone, the researchers analyzed the difference in financial behavior in females who had a male twin compared to females who had a female twin. They found that female twins exposed to more prenatal testosterone (i.e., they had a male twin) were significantly more likely to take financial risk. For example, they were more likely to participate in the stock market, had a higher share of equities or mutual funds in their portfolio, and had a portfolio that exhibited higher volatility. Furthermore, the researchers found that females exposed to higher levels of prenatal testosterone had higher portfolio turnover and had a higher propensity to select lottery type stocks, once again supporting the idea that prenatal testosterone leads to more risk taking.

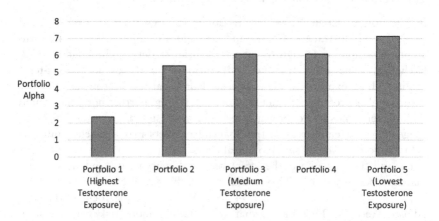

Figure 6.6 Alpha by Testosterone Exposure

The effect of testosterone has also been examined in the context of corporate finance. For example, Ahmed and colleagues used the facial width-to-height ratio of chief executive officers (CEOs) of 104 banks and compared their company's risk taking from 2006 to 2014.[25] The stocks of banks with CEOs who had more masculine faces (i.e., had a higher prenatal exposure to testosterone) had more volatile stock returns and higher idiosyncratic risk. In fact, they found that higher CEO masculinity resulted in an 8 percent increase in the company's stock return. Similarly, Kamiya and colleagues examined facial masculinity in CEOs from 1997 to 2009.[26] They found that more testosterone exposure was positively related to:

- stock return volatility;
- idiosyncratic risk;
- leverage ratio;
- acquisitiveness; and
- the sensitivity of CEO compensation to the stock price.

Overall, these two studies show that CEOs with more masculine facial features tend to exhibit more risk taking in their profession.

Jia and colleagues examined the facial masculinity of 1,136 male CEOs in regard to financial misreporting from 1996 to 2010.[27] They found that CEOs with a higher facial width-to-height ratio (greater exposure to testosterone) were more likely to have an accounting misstatement. In fact, they found that CEOs with above-average facial masculinity had a 98 percent higher likelihood of misreporting on accounting statements.

Testosterone Proxy Validity

The prenatal testosterone proxies may not always be perfect substitutes for the current circulating hormone levels as they are determined prior to birth. The Cognitive Reflection Test (CRT) is a questionnaire designed to measure a person's tendency to carefully contemplate the answer instead of going with a quick intuitive response. The three questions on this test are:

1   A bat and a ball cost $1.10 in total. The bat costs $1.00 more than the ball. How much does the ball cost? _____ cents
2   If it takes 5 machines 5 minutes to make 5 widgets, how long would it take 100 machines to make 100 widgets? _____ minutes
3   In a lake, there is a patch of lily pads. Every day, the patch doubles in size. If it takes 48 days for the patch to cover the entire lake, how long would it take for the patch to cover half of the lake? _____ days

Did you get 5 cents, five minutes, and 47 days? While those are the correct answers, it is very common for people to choose 10 cents, 100 minutes, and 24 days. More correct answers suggest an analytical style of thinking, while fewer correct answers suggest a more intuitive thinking style (See Chapter 12).

Nave and colleagues, and Bosch-Domènech and colleagues, both examined the impact of testosterone on the CRT. Nave and colleagues used a sample of 243 men.[28] Half were given a testosterone cream while the others were given a placebo. They reported that those who were administered exogenous testosterone answered fewer questions correctly. Conversely, Bosch-Domènech and colleagues used a sample of 623 men and women and found that the 2D:4D ratio was related to more correct answers on the CRT.[29] Thus, these proxies may not always be accurate. However, the CRT is not a risk-taking or investment-oriented survey.

Summary

Hormones directly influence the chemical makeup of the brain by altering how neurons send and receive signals throughout the brain. Overall, research consistently finds that testosterone is positively related to risk taking in economic and financial decisions. Conversely, the literature demonstrates that cortisol is related to risk aversion in economic and financial situations. However, it appears that higher levels of testosterone and cortisol are both related to irrational financial decisions such as committing the disposition effect, higher trading frequency, and creating market bubbles and ensuing crashes. Furthermore, using testosterone proxies, research demonstrates that people with more masculine features achieved lower portfolio risk-adjusted returns, higher corporate leverage, and lower financial statement quality.

Questions

1　Why is testosterone known as the male hormone and cortisol known as the stress hormone? How do these hormones influence decisions?
2　How are testosterone and cortisol related to risk aversion?
3　How are testosterone and cortisol related to irrational financial behavior?
4　What do proxies of testosterone exposure tell us about risk aversion?

Notes

1　Thomas Oberlechner and Ashok Nimgade. 2005. Work stress and performance among financial traders. Stress and Health: *Journal of the International Society for the Investigation of Stress* 21:5, 285–293.
2　Uri Gneezy and Jan Potters. 1997. An experiment on risk taking and evaluation periods. *The Quarterly Journal of Economics* 112:2, 631–645.
3　Coren L. Apicella, Anna Dreber, Benjamin Campbell, Peter B. Gray, Moshe Hoffman, and Anthony C. Little. 2008. Testosterone and financial risk preferences. *Evolution and Human Behavior* 29:6, 384–390.

4 Steven J. Stanton, Scott H. Liening, and Oliver C. Schultheiss. 2011. Testosterone is positively associated with risk taking in the Iowa Gambling Task. *Hormones and Behavior* 59:2, 252–256.

5 Steven J. Stanton, O'Dhaniel A. Mullette-Gillman, R. Edward McLaurin, Cynthia M. Kuhn, Kevin S. LaBar, Michael L. Platt, and Scott A. Huettel. 2011. Low-and high-testosterone individuals exhibit decreased aversion to economic risk. *Psychological Science* 22:4, 447–453.

6 Francis J. Greene, Liang Han, Sean Martin, Song Zhang, and Gary Wittert. 2014. Testosterone is associated with self-employment among Australian men. *Economics & Human Biology* 13, 76–84.

7 Jack van Honk, Dennis J.L.G. Schutter, Erno J. Hermans, and Peter Putman. 2003. Low cortisol levels and the balance between punishment sensitivity and reward dependency. *Neuroreport* 14:15, 1993–1996.

8 Narayanan Kandasamy, Ben Hardy, Lionel Page, Markus Schaffner, Johann Graggaber, Andrew S. Powlson, Paul C. Fletcher, Mark Gurnell, and John Coates. 2014. Cortisol shifts financial risk preferences. *Proceedings of the National Academy of Sciences* 111:9, 3608–3613.

9 David Terburg, Barak Morgan, and Jack van Honk. 2009. The testosterone–cortisol ratio: A hormonal marker for proneness to social aggression. *International Journal of Law and Psychiatry* 32, 216–223.

10 Pranjal Mehta and Robert A. Josephs. 2010. Testosterone and cortisol jointly regulate dominance: Evidence for a dual-hormone hypothesis. *Hormones and Behavior* 58, 898–906.

11 Erno J. Hermans, Nick F. Ramsey, and Jack van Honk. 2008. Exogenous testosterone enhances responsiveness to social threat in the neural circuitry of social aggression in humans. *Biological Psychiatry* 63, 263–270.

12 Efrat Barel, Shosh Shahrabani, and Orna Tzischinsky. 2017. Sex hormone/cortisol ratios differentially modulate risk-taking in men and women. *Evolutionary Psychology* 15:1, 1–10.

13 Pranjal H. Mehta, Keith M. Welker, Samuele Zilioli, and Justin M. Carré. 2015. Testosterone and cortisol jointly modulate risk-taking. *Psychoneuroendocrinology* 56, 88–99.

14 John M. Coates and Jessica Kate Herbert. 2008. Endogenous steroids and financial risk-taking on a London trading floor. *Proceedings of the National Academy of Sciences* 105, 6167–6172.

15 John R. Nofsinger, Fernando M. Patterson, and Corey A. Shank. 2018. Decision-making, financial risk aversion, and behavioral biases: The role of testosterone and stress. *Economics & Human Biology* 29, 1–16.

16 John R. Nofsinger, Fernando M. Patterson, and Corey A. Shank. 2019. The Physiology of Investment Biases. Working Paper, University of Alaska, Anchorage.

17 Carlos Cueva, R. Edward Roberts, Tom Spencer, Nisha Rani, Michelle Tempest, Philippe N. Tobler, Joe Herbert, and Aldo Rustichini. 2015. Cortisol and testosterone increase financial risk taking and may destabilize markets. *Scientific Reports* 5:11206, 1–16.

18 Amos Nadler, Peiran Jiao, Cameron J. Johnson, Veronika Alexander, and Paul J. Zak. 2018. The bull of Wall Street: experimental analysis of testosterone and asset trading. *Management Science* 64:9, 4032–4051.

19 Subir Bose, Daniel Ladley, and Xin. 2016. The role of hormones in financial markets. Working Paper, University of Leicester.

20 Ellen Garbarino, Robert Slonim, and Justin Sydnor. 2011. Digit ratios (2D:4D) as predictors of risky decision making for both sexes. *Journal of Risk and Uncertainty* 42:1, 1–26.

21 John V. Nye, Maksym Bryukhanov, Ekaterina Kochergina, Ekaterina Orel, Sergiy Polyachenko, and Maria Yudkevich. 2017. The effects of prenatal testosterone on wages: Evidence from Russia. *Economics & Human Biology* 24, 43–60.

22 John M. Coates, Mark Gurnell, and Also Rustichini. 2009. Second-to-fourth digit ratio predicts success among high-frequency financial traders. *Proceedings of the National Academy of Sciences*, pnas-0810907106.

23 Yan Lu and Melvyn Teo. 2019. Do alpha males deliver alpha? Facial structure and hedge funds. Working Paper, Singapore Management University.

24 Henrik Cronqvist, Alessandro Previtero, Stephan Siegel, and Roderick E. White. 2015. The fetal origins hypothesis in finance: prenatal environment, the gender gap, and investor behavior. *The Review of Financial Studies* 29:3, 739–786.

25 Shaker Ahmed, Jukka Sihvonen, and Sami Vähämaa. 2019. CEO facial masculinity and bank risk-taking. *Personality and Individual Differences* 138, 133–139.

26 Shinichi Kamiya, Y. Han Kim, and Jungwon Suh. 2016. The face of risk: CEO testosterone and risk taking behavior. Working Paper, Nanyang Technological University, Singapore. Retrieved November 20, 2016 from http://papers.ssrn.com/sol3/papers.cfm.

27 Yuping Jia, Laurence van Lent, and Yachang Zeng. 2014. Masculinity, testosterone, and financial misreporting. *Journal of Accounting Research* 52:5, 1195–1246.

28 Gideon Nave, Amos Nadler, David Zava, and Colin Camerer. 2017. Single-dose testosterone administration impairs cognitive reflection in men. *Psychological Science* 28:10, 1398–1407.

29 Antoni Bosch-Domènech, Pablo Brañas-Garza, Antonio M. Espín. 2014. Can exposure to prenatal sex hormones (2D:4D) predict cognitive reflection? *Psychoneuroendocrinology* 43, 1–10.

7   Sleep, Coffee, and Investing

People spend about one-third of their lives sleeping. Sleep is a vital aspect of good decision making. Sleep deprivation is the most severe form of sleep loss. It is a major source of morbidity with widespread health effects, including increased risk of hypertension, diabetes, obesity, heart attack, and stroke. Moreover, sleep deprivation leads to poor decisions that lead to vehicular accidents and medical errors. More spectacularly, a lack of sleep has been linked to the nuclear accident at Chernobyl, Russia, and the destruction of the space shuttle *Challenger*.[1] The health benefits of sleep manifest themselves in many ways. For example, Rodhal argued that sleep is one of the best ways to cope with stress.[2] Similarly, Tanaka and colleagues found that people who have better sleep quality have fewer illnesses and better overall health.[3] In fact, Cappuccio and colleagues found that people who sleep fewer than six hours per night or more than nine hours per night were more likely to have a higher risk of dying earlier.[4] Sleep helps the brain to heal and regenerate itself. During sleep, the brain clears away the toxic by-products of the day's neural activity. A lack of sleep inhibits this process. In fact, Bellesi and colleagues found that chronic sleep loss caused the astrocytes (the cells that regulate the flow of blood through the central nervous system to the neurons for nutrition) to consume portions of the synapses, suggesting that the brain is actually eating itself due to sleep loss.[5] This chapter illustrates how sleep loss influences financial and economic decisions, primarily through changes in risk taking and risk aversion.

Sleep Physiology

Sleep initiates many healing processes in the brain. The lack of sleep inhibits those processes and has significant physical consequences. Poor sleep causes some parts of the brain to be activated less than usual while other parts become overactivated. Did you ever feel like you couldn't think clearly after a miserable night of sleep? This is because sleep deprivation causes a decrease in the activation and neural function of the part of the brain responsible for complex cognitive behavior (the prefrontal cortex and thalamus).[6] Insufficient sleep leads to a general slowing of cognitive response speed and increased variability in performance, particularly for simple measures of alertness, attention, and vigilance. Also, using brain imaging equipment, sleep loss is shown to increase activation in the reward system (nucleus accumbens) while reducing activity in the areas of the brain that regulate emotion (insular cortex) and that facilitate making decisions involving risk and uncertainty (orbitofrontal cortices).[7]

The most severe loss of sleep is called sleep deprivation, which causes the most substantial impacts in the brain. For example, sleep deprivation causes about a 60 percent increase in the magnitude of the activation in the part of the brain that processes emotions (amygdala) and a loss of functional connectivity with the part of the brain that mediates decision making and retrieves remote long-term memory (medial prefrontal cortex), suggesting that different parts of the brain are unable to communicate with each other.[8] Similarly, Nir and colleagues found that parts of the brain not communicating with other parts (in this case the medial temporal lobe) during sleep deprivation resulted in slow behavioral responses (neural lapses).[9] These studies suggest that people experiencing sleep deprivation may have difficulty controlling their emotions, accessing long-term memory, and making decisions.

However, milder forms of sleep loss than sleep deprivation can also reduce cognitive function. Van Dongen and colleagues found that a reduction in sleep by only a few hours, such as only getting six hours of sleep per night, for up to two weeks, provides similar cognitive responses to being totally sleep deprived for 48 hours.[10] Similarly, Bonnet and Arand found that sleep reduction (less sleep than usual) and sleep fragmentation (sleeping and waking up in short intervals) produce similar cognitive deficits as sleep deprivation.[11] Also, Leproult and colleagues found that sleep loss can result in elevated cortisol levels (the stress hormone), which, as we show in Chapter 6, can have an impact on financial decision making.[12]

Finally, sleep helps to consolidate new information into memory during deep sleep (called slow-wave sleep). It also cleans out toxins during rapid eye movement (REM) sleep. Slow-wave sleep is sometimes referred to as sleep-dependent memory processing. Unfortunately, aging is associated with decreased amounts of deep slow-wave sleep and likely impacts the memory processing of older people. REM sleep increases the plasticity in the brain that provides a more favorable condition for the synapses to consolidate the memories in the brain.[13] Sleep also works as a detox for the brain. While you sleep, the space between your brain cells expands by about 60 percent, which removes toxins that have accumulated in the brain throughout the day.[14]

Sleep and Economic Risk Taking

How does a lack of sleep impact decisions that involve economic risk, such as investing? Given the neurological benefits that sleep provide, it is intuitive that sleep should be an important factor of rational financial decision making. A study conducted by Nofsinger and Shank examined the impact of sleep on various types of decisions involving economic risk and uncertainty.

To examine participants' sleep, the researchers asked participants to fill out the Pittsburg Sleep Quality Index. This index is widely known as one of the best surveys to examine sleep and produces an overall index score that is composed of variables for sleep duration, sleep disturbance, sleep latency, day dysfunction, sleep efficiency, self-reported sleep quality, and sleep medication. To assess economic risk aversion, they used the Dynamic Experiments for Estimating Preferences survey to elicit risk and time dependent preferences. The risk decisions ask questions such as "would you prefer gamble A, which is a 50 percent chance to win $100 and a 50 percent chance of losing $15, or gamble B, which is a 90 percent chance to win $40 and a 10 percent chance of winning $10?" In this example, the expected value outcome is much higher for gamble A, which is $42.50 (= 0.5 × $100 + 0.5 × −$15). The expected value of gamble B is only $37 (= 0.9 × $40 + 0.1 × $10). However, gamble A involves the risk of losing money. This risk is not present in gamble B. Thus, this question asks whether you would be willing to give up $5.50 in this situation to guarantee that you do not lose money. A series of these types of questions are asked to assess the person's risk and loss aversion. A series of time decision questions asked questions such as "would you prefer to receive $250 today or $300 in one month?" The authors found that people who scored poorly on the overall sleep quality index were more likely to distort probabilities to be impulsive. That is, they took lower amounts of money sooner rather than wait for a larger outcome later. Of course, this is the opposite of investing for the future. Overall, they found that poor sleep can contribute to irrational financial decisions.[15] They reached several important conclusions:

- Sleep is an essential factor in risk assessment and time preference decisions.
- People increase economic risk taking with poor sleep quality and more sleep disturbances.
- Poor sleep impacts women more than men in time preference decisions.

McKenna and colleagues used a group of participants who had been subjected to sleep deprivation and a group of control participants with normal sleep patterns to conduct an experiment on sleep quality and decisions involving financial risk and uncertainty.[16] These people participated in a lottery-choice task in order to assess their risk and ambiguity preferences. To assess risk, subjects were asked to select between a series of gamble pairs, such as between "Gamble A that has a 67 percent chance of winning $20 and a 33 percent chance of winning $0, or gamble B that has a 33 percent chance of winning $40 and a 67 percent chance of winning $0." To assess ambiguity preferences they chose between "Gamble A which could pay out $20 or $0 but the probabilities are unknown, and gamble B that has a 33 percent chance of winning $40 and a 67 percent chance of winning $0." They found that the people who had been subjected to 23 hours of total sleep deprivation were more willing to take risk when they were considering gambles involving gains but chose less risk when considering gambles with losses. This suggests that sleep deprivation causes people to be loss averse, which is consistent with prospect theory. Additionally, they found no difference between the groups in terms of ambiguity preferences.

Killgore examined the effects of sleep deprivation on economic risk taking. The participants underwent 23 hours of sleep deprivation.[17] These participants then took the Balloon Analog Risk Task (BART), which is an economic test in which the subject earns money every time they push the button to inflate a balloon. There is an unknown threshold whereby, if the subject chooses to inflate the balloon one more time, the balloon will pop, and they will lose all the money they earned in that round. Therefore, every click to inflate the balloon increases the amount of money the subject can earn, but simultaneously increases the risk the balloon will pop, and they will lose all the money earned. The participants took the BART before undergoing sleep deprivation as a baseline and again afterward. Killgore found that participants took less risk in the BART after sleep deprivation suggesting that sleep deprivation causes people to take less risk. Specifically, a risk/benefit ratio was calculated by taking the number of exploded balloons divided by the number of balloons presented (Risk) divided by the total cash received divided by the maximum possible money that could be earned (Benefit):

Risk $\div$ Benefit = (pops/total balloons) $\div$ (total cash/cash possible),

A higher ratio shows a greater risk taken for a lower financial benefit (i.e., more risk). People who were sleep deprived took 21 percent less risk compared to their baseline (Risk/Benefit of 0.76 versus 0.60) when the same participants were not sleep deprived.

Sleep, Economic Risk Taking, and Brain Activation

Chapter 5 demonstrated the importance of neural activity during financial decision making. Given the direct link between sleep and neural function, it is logical to examine brain activation during financial decision making following sleep deprivation.

Venkatraman and colleagues had sleep-deprived participants conduct economic gambles while imaging their brain using functional magnetic resonance imaging (fMRI).[18] The gambles had some interesting deviations that allowed for potential activation in different parts of the brain. First, they allowed the subject to improve the outcome by adding money to a payoff to make the potential gains higher or to make the potential loss smaller. Additionally, after making their choices for a series of gambles (the decision phase), they also saw the results of some of those choices (the outcome phase). Thus, the experiment allowed the authors to examine how sleep deprivation impacts risky economic decisions in both the gain and loss frames, and the potential source of the cognitive behavior from within the brain.

One night of sleep deprivation caused a shift in strategy during the decision phase. Some well-rested participants sought to minimize the effect of the worst loss, as shown by adding money to the largest loss to reduce its magnitude. Sleep deprivation evoked the same people to be less concerned about losses and more concerned about gains by shifting to a strategy that added money that improved the magnitude of the best gain. This change in economic preferences was associated with an increase in ventromedial prefrontal cortex activation as well as a decrease in anterior insula activation during decision making. The ventromedial prefrontal cortex is implicated in the processing of risky decision making and plays a role in regulating emotional responses. Because sleep deprivation caused increases in ventromedial prefrontal cortex activation during the decision-making phase and was also related to an increased focus on the highest gains, Venkatraman and colleagues argue that elevated ventromedial prefrontal cortex activation was predictive of biases placed on the best-ranked outcomes and the large magnitude gains. During decisions that involved a focus on gambles involving a loss, sleep deprivation was associated with decreased right anterior insula activation and decreased dorsomedial prefrontal cortex activation. Activation in these brain regions is believed to be associated with emotional awareness, particularly those relating to negative feelings. Indeed, the study found that activation in the anterior insula was positively related with choices to minimize losses in well-rested participants. However, after sleep deprivation, lowered right anterior insula activation is associated with a decrease in preference for the minimizing loss choices.

Considering both the reduced focus on minimizing losses and the increased focus on maximizing gains together, sleep deprivation appears to create an optimism bias. Subjects behaved as if positive outcomes were more likely and adverse outcomes were less likely. The consequences are that gamblers and traders with gain-seeking, high-risk behavior would also be saddled with an optimism bias and being disproportionately incentivized toward reward seeking while concurrently being desensitized to ongoing losses when sleep deprived.

Sleep and Financial Decisions

Country-wide Analysis

It is common to wake up tired the day after daylight savings time in the spring. If losing an hour of sleep irritates you, you are not alone. Kountouris and Remoundou found that this lack of an hour of sleep caused people to self-report a reduction in their general life satisfaction and mood.[19]

Can this one day of nationwide crankiness affect financial decisions? In aggregate, could these decisions disturb the financial markets? To study the impact of changing the time, Kamstra and colleagues examined stock market returns in the United States, the United Kingdom, and Canada following daylight savings time changes.[20] They found that daylight savings time change caused a significant negative influence on stock returns on the days following daylight savings time changes. The implied loss due to this effect was a loss of approximately $31 billion in market capitalization on the New York Stock Exchange and Nasdaq demonstrating that lack of sleep can cause significant ripple effects throughout the entire stock market.

However, the results from Kamstra and colleagues have been scrutinized because two international crises around daylight savings time complicate the analysis. The authors vigorously defend their conclusions and emphasize that the results came from the time change and not the market's reaction to the crises.[21] In another study, Gregory-Allen and colleagues examined stock returns around daylight savings time changes in 22 stock markets (Australia, Austria, Belgium, Canada, Denmark, Finland, France, Germany, Greece, Hong Kong, Ireland, Italy, Luxembourg, the Netherlands, New Zealand, Portugal, Spain, Sweden, Switzerland, the United Kingdom, and the United States). They found no differences between stock returns on Mondays following daylight savings time changes and other Mondays.[22]

Cai and colleagues used a novel idea to find times when investors are likely to be tired and examined stock returns on those days.[23] Specifically, they examined the impact of sleeplessness due to staying up late at night (after midnight and before 4:00 a.m.) in order to watch the investor's country soccer team play in the World Cup. In addition, they examined the effect of the distraction of matches that occurred during trading hours. The authors found that the effects of both sleeplessness and distraction caused lower stock returns in the country's stock exchange. In fact, the effects were most substantial in countries that were particularly well known for their soccer teams, such as Argentina, Brazil, England, France, Germany, and Japan, as investors in those countries were the most likely to be following their countries' successes in the World Cup.

Individual Focused Experiments

Dickinson and colleagues used an online experiment in two countries to examine the impact of sleepiness on stock returns.[24] Small groups of between seven and 13 people participated in an online experiment at any location of their choice (likely their home). The experiment took two hours and required them to be online at precise times of the day at their local time which was either noon, 4:00 p.m., 8.00 p.m., midnight, 4:00 a.m., or 8:00 a.m. The subjects were recruited from the United States and New Zealand, which are 16 time zones apart. The subjects were endowed with $50 and six shares of the experimental stock. The participants then engaged in a stock trading game by trading with each other. The subjects participated in either the low-return treatment or the high-return treatment. In the low-return treatment, the person received 10 percent in interest on cash held and either a $0.40 or $1.00 dividend per share on stock held. In the high-return treatment, the investor received 20 percent interest on cash held and a dividend of either $1.10 or $1.70 per share on the stock. The expected value of the stock in each treatment is $7. This is computed as the average of each dividend divided by the interest rate yields: low-return treatment $7 = ($0.40 + $1.00 ÷ 2) ÷ 0.10 and high-return treatment $7 = ($1.10 + $1.00) ÷ 2) ÷ 0.20. However, the demand for buying and selling of the stock often caused the price to substantially deviate from the $7 value. The trading game went through 30 rounds, with 15 rounds each of the low-return and high-return treatments.

Eight groups of traders were all located in the same time zone (dubbed a "local market"). Another 12 groups had people from both time zones, 16 hours apart (dubbed a "global market"). A local market game included all participants playing at either standard times of the day or all at night. The authors supplemented the time of day with a survey question about how sleepy the participants felt. The global market games had a mix of people playing from work during both day and night times. The study examined several trading results: the occurrence of stock price bubble deviations from the $7 stock value; the riskiness of the portfolios held; the price difference between offers to buy/sell and the current price; and trader earnings. They made the following conclusions:

- Price bubbles were more substantial and lasted longer for global markets, which had tired but alert participants.
- Compared to alert participants, sleepy participants held riskier portfolios and held them later in the game, which was riskier when the stock was selling over the $7 value.
- Tired participants exhibited a form of price extremism in which they offered bid and ask prices with higher deviations from the current price trade.
- No difference was found between tired and alert participant investment earnings.

Sleep Deprivation, Stimulants and Risk Taking

When most people wake up tired, they typically reach for their coffee or tea, which may explain why caffeine is the most consumed drug in the world. Other stimulants that some people use to stay awake are dextroamphetamine and modafinil. Dextroamphetamine is a controlled substance that is frequently prescribed for attention deficit hyperactivity disorder (ADHD) and narcolepsy but has also been used by the military to give to soldiers to prevent fatigue during night-time missions. Modafinil is frequently prescribed for narcolepsy, shift work sleep disorder, or obstructive sleep apnea. Taking these stimulants makes you feel more alert, but do they help to overcome the neurological detriments of poor sleep for better decision making?

To examine the impact of stimulants, Killgore and colleagues had participants undergo 44 hours of continuous wakefulness and then ingest either

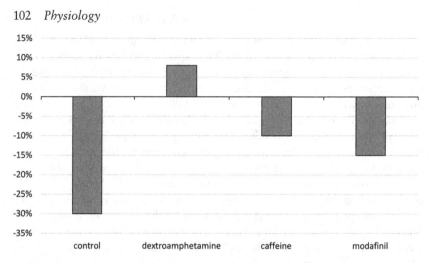

Figure 7.1 Sleep Deprivation, Stimulants, and the BART – Change in Risk-Taking

caffeine, dextroamphetamine, modafinil, or a placebo.[25] Two hours after ingesting the stimulant or placebo, the participants completed the BART (discussed above). The authors found that after sleep deprivation, participants had a decline in risk taking compared to their baseline risk-taking levels. An exception was those who ingested dextroamphetamine. Figure 7.1 demonstrates that from baseline levels people who were given a placebo took about 30 percent less risk, while people who took dextroamphetamine took nearly 10 percent more risk. The other stimulants provided a negative change from the baseline; however, it was not as severe as the placebo group. These results suggest that caffeine has some ability to restore the cognitive process for evaluating risk. For prescription stimulants, dextroamphetamine may be effective for people who do not wish their sleep patterns to influence their financial decision making.

Killgore and colleagues conducted a similar experiment in which they had participants undergo 44 hours of sleep deprivation and then take either caffeine, dextroamphetamine, modafinil, or a placebo. Following this they were asked to perform the Iowa Gambling Task.[26] For this task the subjects must select a card from one of four card decks (A, B, C, and D). Each card causes them either to win or lose money. Decks A and B have a 50 percent probability of winning $100 and a 50 percent probability of losing $250. Decks C and D have a 50 percent probability of winning $50 and a 50 percent probability of losing $50. The subjects were not told of this distribution. This task tests whether subjects have the neural processes available to realize that decks A and B are disadvantageous and thus avoid selecting from them. The sooner they realize this, so much the better for them. In contrast to the results using the BART, the authors found that in the Iowa Gambling Task the stimulants provided no improvement in the ability to adapt and stop selecting from the disadvantageous decks. Figure 7.2 shows good deck minus bad deck choices.

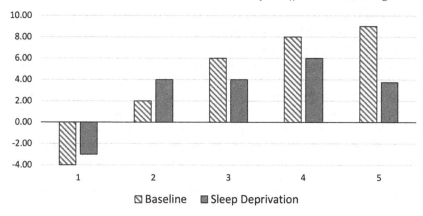

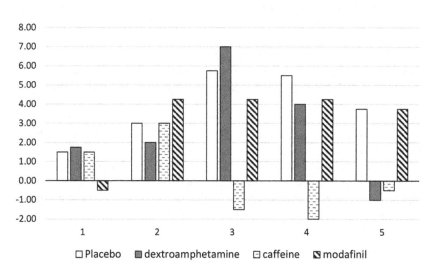

Figure 7.2 Sleep Deprivation, Stimulants, and the Iowa Gambling Task – Good Minus Bad Deck Choices

We can see that as time goes on, participants became better at realizing that there are good decks vs. bad decks; however, the sleep-deprived group had a harder time distinguishing between the decks. When looking at the impact of stimulants, the graph shows that depending on the round, participants taking different stimulants or the placebo selected more cards from the good deck versus the bad deck. This shows that there is no significant impact between the placebo and the stimulants for cognition.

Most studies investigating sleep and financial decision making have subjects stay awake and then conduct the experiment. Thus, increments of sleep loss are rarely studied. In this vein, Killgore and colleagues examined to what extent impaired decision making as a result of sleep deprivation was exacerbated by increasing sleepiness.[27] In addition, they studied how the effect can be reversed by caffeine. Participants completed the Iowa Gambling Task before the experiment to determine a baseline. They retook the Iowa Gambling Task after 51 hours of sleep deprivation and again after 75 hours of sleep deprivation. Additionally, 12 of the subjects received a dose of caffeine every two hours each night. The authors' results showed that before undergoing sleep deprivation, the participants quickly learned to avoid the disadvantageous decks. After sleep deprivation, the same participants drew from the disadvantageous deck more frequently; thus, the sleep loss impacted their cognitive process. They also found that the severity of impaired decision making did not increase from the level of participants having 51 hours of sleep deprivation to the level of those suffering 75 hours of sleep deprivation. Thus, it appears that cognitive degradation occurred with sleep loss, but was not necessarily varied with the extent of the sleep loss. Possibly more importantly, caffeine did not have a significant impact on improved decision making.

Summary

Lack of sleep or poor-quality sleep can directly influence the neurological function of the brain through decreased activation and neural function. Overall, the research shows that people who are sleep deprived or who lack quality sleep are both risk and loss averse. Additionally, they make more irrational financial decisions, as shown in studies using the Iowa Gambling Task. Furthermore, it does not appear that stimulants such as caffeine can reverse the impact of sleep deprivation, although it has been shown to marginally improve cognition in certain contexts. Thus, stimulants do not appear to be the answer to tired financial professionals, especially given the potential side effects.

Questions

1 How does a period of more than 24 hours without sleep or alternatively getting consistently less than six hours of sleep affect the body?
2 How does poor sleep affect risk taking?
3 What are two ways that researchers examine the effect of sleep on the aggregate stock market?
4 How do stimulants like coffee affect financial decisions?

Notes

1 Merrill M. Mitler, Mary A. Carskadon, Charles A. Czeisler, William C. Dement, David F. Dinges, and R. Curtis Graeber. 1988. Catastrophes, sleep, and public policy: Consensus report. *Sleep* 11:1, 100–109.
2 K. Rodahl. 2003. Occupational health conditions in extreme environments. *Annals of Occupational Hygiene* 47:3, 241–252.

3 Hideki Tanaka, Kazuhiko Taira, Masashi Arakawa, Atushi Masuda, Yukari Yamamoto, Yoko Komoda, Hathuko Kadegaru, and Shuichiro Shirakawa. 2002. An examination of sleep health, lifestyle and mental health in junior high school students. *Psychiatry and Clinical Neurosciences* 56:3, 235–236.

4 Francesco P. Cappuccio, Lanfranco D'Elia, Pasquale Strazzullo, and Michelle A. Miller. 2010. Sleep duration and all-cause mortality: A systematic review and meta-analysis of prospective studies. *Sleep* 33:5, 585–592.

5 Michele Bellesi, Luisa de Vivo, Mattia Chini, Francesca Gilli, Giulio Tononi, and Chiara Cirelli. 2017. Sleep loss promotes astrocytic phagocytosis and microglial activation in mouse cerebral cortex. *Journal of Neuroscience* 37:21, 5263–5273.

6 Sean P.A. Drummond, Gregory G. Brown, John L. Stricker, Richard B. Buxton, Eric C. Wong, and J. Christian Gillin. 1999. Sleep deprivation-induced reduction in cortical functional response to serial subtraction. *Neuroreport* 10:18, 3745–3748; Maria L. Thomas, Helen C. Sing, Gregory Belenky, Henry H. Holcomb, Helen S. Mayberg, Robert F. Dannals, Henry N. Wagner, Jr., David R. Thorne, Kathryn A. Popp, Laura M. Rowland, Amy B. Welsh, Sharon M. Balwinski, and Daniel P. Redmond. 2000. Neural basis of alertness and cognitive performance impairments during sleepiness. I. Effects of 24 h of sleep deprivation on waking human regional brain activity. *Journal of Sleep Research* 9:4, 335–352; Jim Horne. 2012. Working throughout the night: Beyond "sleepiness"—impairments to critical decision making. *Neuroscience & Biobehavioral Reviews* 36:10, 2226–2231.

7 Vinod Venkatraman, Y.M. Lisa Chuah, Scott A. Huettel, and Michael W.L. Chee. 2007. Sleep deprivation elevates expectation of gains and attenuates response to losses following risky decisions. *Sleep* 30:5, 603–609.

8 Seung-Schik Yoo, Ninad Gujar, Peter Hu, Ferenc A. Jolesz, and Matthew P. Walker. 2007. The human emotional brain without sleep: A prefrontal amygdala disconnect. *Current Biology* 17:20, R877–R878.

9 Yuval Nir, Thomas Andrillon, Amit Marmelshtein, Nanthia Suthana, Chiara Cirelli, Giulio Tononi, and Itzhak Fried. 2017. Selective neuronal lapses precede human cognitive lapses following sleep deprivation. *Nature Medicine* 23:12, 1474–1480.

10 Hans P. A. Van Dongen, Greg Maislin, Janet M. Mullington, and David F. Dinges. 2003. The cumulative cost of additional wakefulness: Dose-response effects on neurobehavioral functions and sleep physiology from chronic sleep restriction and total sleep deprivation. *Sleep* 26:2, 117–126.

11 Michael H. Bonnet and Donna L. Arand. 2003. Clinical effects of sleep fragmentation versus sleep deprivation. *Sleep Medicine Reviews* 7:4, 297–310.

12 Rachel Leproult, Georges Copinschi, Orfeu Buxton, and Eve Van Cauter. 1997. Sleep loss results in an elevation of cortisol levels the next evening. *Sleep* 20:10, 865–870.

13 Susanne Diekelmann and Jan Born. 2010. The memory function of sleep. *Nature Reviews Neuroscience* 11:2, 114–126.

14 Lulu Xie, Hongyi Kang, Qiwu Xu, Michael J. Chen, Yonghong Liao, Meenak-shisundaram Thiyagarajan, John O'Donnell, Daniel J. Christensen, Charles Nicholson, Jeffrey J. Iliff, Takahiro Takano, Rashid Deane, and Maiken Neder-gaard. 2013. Sleep drives metabolite clearance from the adult brain. *Science* 342:6156, 373–377.

15 John R. Nofsinger and Corey A. Shank. 2019. DEEP sleep: The impact of sleep on financial risk taking. *Review of Financial Economics* 37:1, 92–105.

16 Benjamin S. McKenna, David L. Dickinson, Henry J. Orff, and Sean P.A. Drummond. 2007. The effects of one night of sleep deprivation on known-risk and ambiguous-risk decisions. *Journal of Sleep Research* 16:3, 245–252.

17 William D. Killgore. 2007. Effects of sleep deprivation and morningness-eveningness traits on risk-taking. *Psychological Reports* 100:2, 613–626.

18 Vinod Venkatraman, Scott A. Huettel, Lisa Y.M. Chuah, John W. Payne, and Michael W.L. Chee. 2011. Sleep deprivation biases the neural mechanisms underlying economic preferences. *Journal of Neuroscience* 31:10, 3712–3718.

19 Yiannis Kountouris and Kyriaki Remoundou. 2014. About time: Daylight saving time transition and individual well-being. *Economics Letters* 122:1, 100–103.

20 Mark J. Kamstra, Lisa A. Kramer, and Maurice D. Levi. 2000. Losing sleep at the market: The daylight saving anomaly. *American Economic Review* 90:4, 1005–1011.

21 J. Michael Pinegar. 2002. Losing sleep at the market: The daylight saving anomaly: Comment. *American Economic Review* 92:4, 1251–1256; Mark J. Kamstra, Lisa A. Kramer, and Maurice D. Levi. 2002. Losing sleep at the market: The daylight saving anomaly: A reply. *American Economic Review* 92:4, 1257–1263.

22 R. Gregory-Allen, B. Jacobsen, and W. Marquering. 2010. The daylight saving time anomaly in stock returns: Fact or fiction? *Journal of Financial Research* 33:4, 403–427.

23 Jinghan Cai, Manyi Fan, Chiu Yu Ko, Marco Richione, and Natalie Russo. 2018. Physiology, Psychology, and stock market performance: Evidence from sleeplessness and distraction in the World Cup. Working Paper, University of Scranton, July.

24 David L. Dickinson, Ananish Chaudhuri, and Ryan Greenaway-McGrevy. Forthcoming. Trading while sleepy? Circadian mismatch and excess volatility in a global experimental asset market. *Experimental Economics*.

25 William D.S. Killgore, Nancy L. Grugle, Desiree B. Killgore, Vrain P. Leavitt, George I. Watlington, Shanelle McNair, and Thomas J. Balkin. 2008. Restoration of risk-propensity during sleep deprivation: Caffeine, dextroamphetamine, and modafinil. *Aviation, Space, and Environmental Medicine* 79:9, 867–874.

26 William D.S. Killgore, Nancy L. Grugle, and Thomas J. Balkin. 2012. Gambling when sleep deprived: Don't bet on stimulants. *Chronobiology International* 29:1, 43–54.

27 William D.S. Killgore, Erica L. Lipizzi, Gary H. Kamimori, and Thomas J. Balkin. 2007. Caffeine effects on risky decision making after 75 hours of sleep deprivation. *Aviation, Space, and Environmental Medicine* 78:10, 957–962.

8 How Wellness Influences Financial Decisions

The Centers for Disease Control and Prevention (CDC) reported that more than 70 percent of Americans were either overweight or obese in 2016.[1] Additionally, according to the Organization for Economic Co-operation (OECD), the United States is the most obese country in the world.[2] Figure 8.1 illustrates the number of Americans who are either overweight or obese. In 2000, 34 percent of Americans were overweight, and 30 percent were obese. In 2016, the number of people who were overweight decreased to 31 percent, but the number of Americans who were obese increased to 40 percent. Finkelstein and colleagues estimated that the medical costs related to obesity alone were approximately $147 billion in 2008, a figure that has undoubtedly increased.[3] Hammond and Levine summarized several academic papers and found that obese people spend between 36 percent to 100 percent more on direct medical costs per year than do people of healthy weight. Obese students miss approximately two more days per month of school, which results in about 0.2 fewer grades completed.[4] The authors also found that obese people miss about 1.5 more days of work per month due to their health, which resulted in a loss of up to $6 billion per year to companies. Guthrie and Sokolowsky examined the impact of obesity on financial distress. The authors found that debt delinquency was 20 percent higher for obese people versus people of healthy weight. Furthermore, they found that poor health and an increase in medical expenditure partially explained the increase in delinquent debt.[5] Low exercise and a poor diet can contribute to poor health. Aerobic exercise and a proper diet have direct positive effects on cognition and mood. However, does obesity, poor health, diet, and exercise affect financial decisions?

Health and Obesity

There are two different lines of thinking about how health and obesity can affect financial decisions. On the one hand, people in poor health are more likely to keep money in liquid assets so that they can pay for their future medical bills. An alternative theory illustrates that people in poor health or those struggling with their weight go through experiences in life that

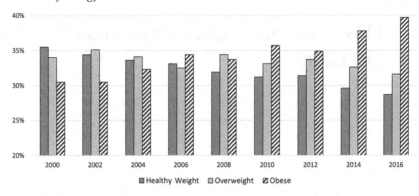

Figure 8.1 Prevalence of Obesity in the United States

influence their decisions. For example, Conti and Heckman suggested that social conditions during childhood are negatively related to weight.[6] Similarly, Cairney and colleagues found that social stigma was a contributor to an obese person's anxiety.[7] In fact, Vartanian stated that obese people were discriminated against in the same manner as drug addicts.[8] These earlier experiences may lead to people who are in poor health and obese becoming less optimistic and having a higher risk of depression.[9] Given that research shows how mood influences financial decisions, it is likely that people in poor health will make mood-impacted financial decisions (See Chapter 9 on emotions and mood).

Exercise and Diet

Diet and exercise are factors of health status, both in relation to obesity and independently. Specifically, it has long been known that poor dietary choices and a lack of consistent exercise are common among obese people.[10] In addition, diet and exercise have been shown to impact regions of the brain that control decision making. A diet high in saturated fat and refined sugar can negatively influence brain function in such a manner that it inhibits learning and memory,[11] while vegetable and fruit consumption have a positive effect on cognitive function.[12] Exercise also elicits positive reactions in the brain. Exercise increases brain volume in the regions related to emotions, problem solving, memory, risk taking, and decision making.[13] Consequently, exercise increases cognitive function.[14] Both exercise and improved diet can improve mood,[15] which suggests that they can influence peoples' financial risk-taking appetite (see Chapter 9). Thus, a lack of exercise and a poor diet can negatively impact financial decision making through poor health and obesity. In addition, aerobic exercise and improved diet can directly and positively impact financial decision making.

Physical Health and Financial Decisions

Health and Portfolio Choice

The basic financial theory is that over the long term taking systematic risk leads to earning risk premiums, which provide higher returns. People who take higher investment risks during their lifetimes can grow their investment portfolios substantially larger. Thus, if unhealthy people tend to take less investment risk, they build smaller portfolios during their lifetimes which negatively impacts their retirement lifestyle.

Rosen and Wu examined portfolio decisions using the Health and Retirement Study (HRS) that follows about 7,000 households with heads of household aged between 51 and 61 years.[16] The study used the surveys conducted in 1992, 1994, 1996, and 1998. Figure 8.2 illustrates how health status impacts portfolio choice. They found that healthy married couples were most likely to have retirement accounts (such as Individual Retirement Accounts or 401ks), individual bonds, and risky individual assets like stocks and mutual funds. Conversely, single and unhealthy individuals were least likely to hold all three of these assets.

Edwards developed an economic model to illustrate the relationship between health and portfolio choice when health becomes poor.[17] His model shows that adverse shocks to a person's health cause concern about higher health expenditures and potentially lower income from missed work. Given this shock to their health, the rational choice is to lower the amount of risk in their portfolio so that people can be more confident about money being available in the short term. Similarly, other research connects declining health associated with aging to shifts in increased risk aversion associated with portfolio choice.[18]

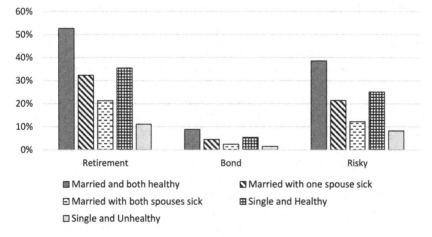

Figure 8.2 Health and Asset Holdings

Goldman and Maestas examined how medical expenditure risk affected household portfolio choice using Medicare supplemental health insurance data.[19] The authors examined Medicare supplemental health insurance because Medicare beneficiaries had a risk of large out-of-pocket medical costs and the supplemental insurance policies helped to fill this gap. The authors found that people who held supplemental policies held 7.1 percent more risky assets than those who did not hold the supplemental insurance policy. This result suggests that retirees are more likely to hold risky assets if they can lower their risk of large medical bills.

It appears that the relationship between health and portfolio choice is not monotonic. Bressan and colleagues examined the impact of health on portfolio choice using the Survey of Health, Ageing, and Retirement in Europe (SHARE).[20] This study categorized people into five health quality groups: poor, fair, good, very good, and excellent. Overall, the authors found that only those in the poor health group negatively affected the choice of a risky portfolio. People in excellent, very good, good, and fair health all held similar portfolios.

In addition, the relationship between health and portfolio risk taking is complicated by other factors associated with poor health like depression. Addoum and colleagues examined the impact that obesity had on portfolio choice using retirement data in the United States and Europe.[21] They found that people who were overweight were less likely to participate in the stock market, and if they did they held riskier assets due to being less optimistic. However, the authors found that the results were primarily driven by education, cognitive ability, and race.

There is also contrary evidence. Love and Smith used the HRS data from 1992 to 2006 to examine whether health caused portfolio choice or if asset allocation was influenced by other characteristics such as risk attitudes, impatience, information, and motivation.[22] After accounting for these characteristics, the authors found that health did not affect portfolio choice in single households and had only a small effect on married households.

Health and Risk Taking: Experimental Evidence

Do unhealthy or obese people exhibit fundamental differences in financial risk tolerance compared to healthy people? People with higher health risks may have to invest in less risky assets due to liquidity needs. However, how do they behave during risky gambles? That is, if they are given two possible outcomes, do they still take lower risk outside of their investment portfolios?

Patterson and Shank employed the Dynamic Experiments for Estimating Preferences survey to elicit risk and time decision preferences to determine how health affected financial decisions.[23] In this experiment, participants were asked questions such as "Would you prefer gamble A, which is a 50 percent chance to win $100 and a 50 percent chance of losing $15, or gamble B, which is a 90 percent chance to win $40 and a 10 percent chance of winning $10?" For this question, gamble A is clearly riskier because it involves the potential for losing money while gamble B does not. However, gamble A also offers a higher expected return of $42.50 (= 0.5 × $100 + 0.50 × −$15) then the expected payoff of $37.00 (= 0.9 × $40 + 0.10 × $10) for gamble B. The survey offered many questions like this one, which allowed the scholars to focus on the participants' risk aversion level. There were also time decision questions in the survey. These questions asked questions such as "Would you prefer to receive $250 today or $300 in one month?" Note that waiting one month offers a 20 percent return (= ($300–$250)/$250), which is a nearly an 800 percent annual rate of return. The participants answered several such questions to focus on their time preferences. The participants were also given a survey about their personal health status. The authors' results can be summarized as follows:

- A higher body mass index (BMI), was related to higher risk and loss aversion.
- A higher BMI was related to a higher discounting rate, which suggests that they were more impatient and less likely to delay current gratification for future wealth.
- People who ate more vegetables were more likely to delay current gratification for future wealth.
- People who ate more fruit and exercised vigorously were less likely to be loss averse.

Employee Health and Company Performance

Thus far, we have investigated how a person's health impacts their financial decision making. However, in aggregate, the health of a community might impact the productivity and profitability of the local firms and the economy. That is, communities that have a greater population of people in poor health may impact the local companies because those people were employees of the firms.

Nofsinger and Shank used the Behavioral Risk Factor Surveillance System (BRFSS) surveys collected by CDC to examine how community health affected the stock returns of local companies.[24] The BRFSS collected interviews conducted with thousands of people across the United States and asked them questions about their overall health, how many days per month they were in either poor physical or mental health, their height, and weight, etc. The authors found that companies with headquarters in areas with higher rates of obesity and higher rates of people self-reporting more days of poor mental or physical health experienced lower stock returns. The authors argued that people in poor health were less productive due to absenteeism and presenteeism, which caused companies to report lower profits and stock returns. Absenteeism is the loss of productivity due to missed days of work, while presenteeism is low productivity while at work.

Following this same line of thinking, Agrawal and Lim examined the impact of obesity and corporate policies.[25] The authors found that firms headquartered in areas with higher obesity rates tended to invest less in tangible assets and research and development, had lower sales and asset growth, and had lower profits measured by return on assets. Overall, the authors found that the demographics of the local community influenced the firm's corporate policies.

Edmans studied a similar question and examined the impact of employee satisfaction on stock prices.[26] Edmans created portfolios each year that comprised all the firms that appear in *The 100 Best Companies to Work For in America*, a book that was first produced each year by Levering, Moskowitz, and Katz and which later featured in *Fortune* magazine every January. The list of 100 best companies to work for was created by using an employee questionnaire that covered topics such as attitudes toward management, job satisfaction, fairness, and camaraderie and comes from company information such as demographic composition, pay and benefits, and company culture. In a Ted Talk in 2015, Edmans used the example of Costco stating that the company paid its employees $20 per hour, which was nearly double the industry average of $11 per hour. In addition to the high pay, the company provided 90 percent of its employees' health insurance.[27] Some analysts believe that the more employees are paid, the less the company will profit. However, Edmans concluded that treating employees well and caring about their well-being created more firm value. His overall conclusion was that companies that made this Best Companies list experienced higher returns than their peers by 2–3 percent per year and argued that employee satisfaction was positively related to shareholder return.

Medical research shows that exercise moderates stress and can improve cognitive function and mood. As such, Limbach and Sonnenburg examined whether having a CEO who exercised was beneficial to the firm.[28] They compared firms with a CEO who had competed in endurance sports such as marathons (e.g., Jack Brennan from Vanguard, Robert Iger from Walt Disney, Klaus Kleinfeld from Alcoa, John Legere from T-Mobile, or Steven Reinemund from PepsiCo) versus companies who had a CEO who had not. After controlling for factors such as governance characteristics, firm characteristics, and past performance, these authors found that firms who had a "fit" CEO had higher profitability as measured by Tobin's Q. Tobin's Q is a common measure of profitability that compares the market value of the firm to the replacement costs of its assets.

Society Illness and Stock Market Performance

How do you feel when you are sick with the flu? During the flu season, millions of people across the United States are plagued with this illness. As such, it is intuitive that possibly having millions of investors sick with the flu rather than working would have an effect on stock returns.

McTier and colleagues examined how the flu season impacted stock trading activity on the New York Stock Exchange.[29] Using weekly observations of flu occurrences in regions in the United States from CDC, they investigated whether large numbers of people being sick impacted the stock market. In general, increases in flu incidence were associated with decreased trading volume, lower realized volatility, widened bid-ask spreads, and lower returns.

For individual companies, a flu incidence near their headquarters was associated with a change in their stocks' trading activity. Thus, it appears that the effect is not due to lost employee productivity or lower consumption in the region. The lower volume and volatility effects were more closely tied to New York area flu incidence. This is consistent with the flu impacting the market makers of the exchange and the trading of institutional investors. The lower volatility result was consistent with greater absenteeism of these large investors and market makers and the associated reduction of their information production. The lower returns were tied to flu incidence nationwide. This likely reflects expectations about declined real economic activity.

Mental Health and Financial Decisions

It has been estimated that close to 30 percent of the U.S. population experiences at least one mental or substance abuse disorder each year. Common psychiatric conditions include depression, anxiety, phobias, alcoholism, and obsessive-compulsive disorder. Poor mental health could affect investment and portfolio choices for several reasons. Mental health may alter cognitive abilities (see Chapter 12) and affect one's ability to regulate mood and emotion (see Chapter 9). Mental health shocks may also alter a person's degree of risk aversion and discount rate, which impacts motivation to invest for the future. Finally, mental health disorders reduce available funds for investing through lower productivity and increase medical spending. The result may be that poor mental health causes a household to allocate more of its portfolio toward more liquid assets, like cash. Also, people with mental health issues that affect emotions or mood may have more difficulties evaluating investment opportunities.

Bogan and Fertig investigated whether mental health status could help to explain household portfolio allocation.[30] The study used data from seven biennial waves of the HRS. The surveys collected standard demographic information as well as information about financial assets and detailed physical and mental health conditions. For this study, mental health problems included depression, anxiety, phobias, alcoholism, and obsessive-compulsive disorder. Did these mental health afflictions affect investment behavior and portfolio optimization? The study found that households suffering with mental illness decreased their investments in risky investments. Single women diagnosed with psychological disorders were found to have an increased probability of holding cash-type assets and an increased allocation to those safe investments. Men with cognitive functioning problems increased financial assets devoted to retirement and pension accounts. Note that mental health issues affected males, females, and households differently.

Bogan and Fertig expanded their previous work by exploring how psychological distress impacted retirement savings account behavior.[31] Mental health problems could impact retirement savings when they impact risk preferences or intertemporal discounting. Indeed, households affected by mental health problems were more risk averse, which could decrease allocations to risky investments like stocks. Also, mental health issues can affect time preferences. Typically, poor mental health was associated with a higher discount rate, which focused the person on the present at the expense of the future.

The study used the Panel Study of Income Dynamics and the HRS, which were both biennial surveys. The sample was limited to men and women below retirement age (65 years for men and 67 for women). The study examined the relation between incidences of mental health distress and retirement plan choices. They found that psychological distress had a large and economically significant effect on retirement plan behavior. Specifically, psychological distress was associated with:

- a significantly decreased probability of holding retirement accounts by as much as 24 percent;
- a decreased share of financial assets devoted to retirement accounts by as much as 67 percent; and
- increased withdrawals from retirement accounts for married individuals.

Lindeboom and Melnychuk analyzed the impact of mental health problems on the decision to invest in risky financial assets in 11 European countries.[32] They used data from the SHARE. Specifically, the study investigated the association between depression and the holding of risky financial assets. Depression may distort a person's perceptions and the way they interpret information during decision making. Depression causes people to fear taking risks, which in turn causes an increase in their risk aversion. Suffering depression can, therefore, cause people to evaluate investment opportunities from a more pessimistic frame in the assessment of the probabilities of gains and losses. The study reported that suffering symptoms of depression lowered the probability of a person acquiring risky financial assets, such as stocks. Thus, depressed people experience lower portfolio investment rates and lower future wealth.

Impact of Finance on Health

So far, we have documented how a person's health can influence investment returns and policy. However, can the opposite be true? Can the financial environment affect a person's health? Ong and colleagues examined the effects of chronic debt on behavior using the unanticipated debt relief program in Singapore. This relief program provided low-income families mired in debt with up to $5,000, which was equivalent to about three months' worth of household income. They found that having their debt accounts paid off decreased the likelihood of anxiety by 11 percent and increased cognitive function.[33]

Furthermore, Deaton examined the well-being of Americans during the financial crisis from 2008 to 2009. During the crisis, the unemployment rate rose from 4.8 percent to 10.6 percent, which resulted in a drop in disposable income of 1.7 percent. This had an adverse effect on people's health. Even if a job loss didn't directly impact a person, the effects of the poor economy had an impact on everyone's thoughts. Deaton found that during the collapse of the stock market, Americans reported a large decline in self-reported well-being and large increases in worry, stress, and anxiety as reported in the Gallup Healthways Well-being Index generated from daily surveys of 1,000 randomly sampled people.[34] Similarly, McInerney and colleagues examined the impact that the 2008 stock market crash had on mental health in older Americans using the HRS. The authors found that there was a positive relationship between the amount of wealth lost due to the crash and the increased feeling of depression and the use of antidepressant drugs.[35]

Schwandt also examined the impact of wealth shocks on health outcomes in retirees using the HRS. In this paper, the author used direct measures of how much stock the retirees held.[36] The results showed that a 10 percent increase in wealth led to a 3 percent increase in physical health and mental health quality and survival rates. Ratcliffe and Taylor examined the same question in the United Kingdom using data from the British Household Panel Survey from 1991 to 2008.[37] The authors reported a similar result showing that stock declines were related to poor mental health. Additionally, they found that increased stock market volatility also increased poor mental health.

Cotti and colleagues examined how large changes in the Dow Jones Industrial Average stock index impacted health using the BRFSS with a focus on the sharp declines in the stock market in 1987 and 2008–2009.[38] The authors found that large declines in the stock market were associated with worsening self-reported mental health. Furthermore, they also found that during episodes of stock market decline it was reported that people smoked more often, had more binge drinking episodes, and had more fatal car accidents that involved alcohol. Overall, these results suggest that people try to cope with their depression caused by stock market declines by participating in risky behavior such as smoking and drinking. Similarly, Giulietti and colleagues also found that there were more fatal car crashes following days of poor stock returns.[39]

However, financial distress does not only affect adults. Cotti and Simon examined the impact of the stock market on the well-being of children.[40] They found that the stock market crash in 2008 resulted in more child hospitalizations, sick days from school, and lower overall health quality. These results suggest that the increased economic stress on adults also spills over to children.

Similarly, Engelberg and Parsons used individual patient records from all the hospitals in California from 1983 to 2011 to examine the correlation between admissions and stock returns.[41] The authors found that on days following a large stock decline, the number of patients admitted for psychological conditions such as anxiety, panic disorder, or major depression increased. This suggests that the stock market's poor performance causes physical stress on people throughout the country. However, research suggests that the effects of depression caused by stock declines goes beyond smoking and drinking. Wisniewski and Lambe found that stock declines led to an increase in suicide rates, suggesting that there should be an increase in suicide prevention strategies during poor economic times.[42]

Kalcheva and colleagues examined the impact of economic policy uncertainty on healthy choices using the economic policy uncertainty index calculated from policyuncertainty.com.[43] This index was created based on news, which examined the number of times words such as "uncertainty" or "uncertain" appeared in economic or political articles, disagreement on forecasts for consumer price index, disagreement on federal/state/local governmental purchases, and the discounted dollar-weighted tax code. After controlling for the effects of stock market changes, the authors found that economic policy uncertainty increased impulsive behavior such as drinking, alcoholic beverage consumption, binge drinking, and smoking. These results suggest that it may be the uncertainty of the future that is causing poor health choices rather than the declines in the stock market.

Furthermore, financial markets do not only affect investors. Kahn and Cooper examined the mental health of 225 dealers and traders working at the London Stock Exchange.[44] They found that the professional traders exhibited higher levels of stress and anxiety than the general population and that many of them seemed to cope with this stress through alcohol. Additionally, Kahn and colleagues examined whether the stress of being a financial trader was different on Wall Street compared to the London Stock Exchange and found similar levels of stress and alcohol use.[45]

Finally, Oberlechner and Nimgade examined work stress and performance among 325 financial traders.[46] They found that 32 percent of the traders reported either "very high" or "extremely high" stress levels. The most significant stressor for the financial traders was the focus on the profit goal followed closely by long working hours.

Summary

Health and finance are two areas that appear to be uniquely intertwined. The literature documents a rational response of people in poor health being more risk averse since it is likely that they will need their money to cover their medical expenditure. Those in very poor health drive this result. Low exercise levels and a poor diet contribute to obesity and poor health, which impact investment decisions. Aerobic exercise and improved diet had direct positive effects on cognition and decision making. Mental health conditions also have important financial impacts, although it appears to result in different financial behaviors between men, women, and households. Health also spills over into corporations as research shows that companies headquartered in communities with poor health have lower stock returns, less investment in tangible assets and research and development, and lower growth rates. Conversely, companies that treat their employees well and truly care about their well-being tend to achieve 3 percent higher returns per year compared to their peers. Finally, economic downturns have an inverse relationship with a person's mental well-being and healthy choices.

Questions

1   How does health status influence portfolio choice? Is this a rational decision?
2   How does employee health impact firm performance?
3   How does a person's health impact financial decisions?
4   How can financial markets influence a person's health?

Notes

1   NCHS, National Health and Nutrition Examination Survey. See Appendix I, National Health and Nutrition Examination Survey (NHANES).
2   OECD. 2017. OECD Health Statistics 2017.
3   Eric A. Finkelstein, Justin G. Trogdon, Joel W. Cohen, and William Dietz. 2009. Annual medical spending attributable to obesity: Payer-and service-specific estimates. *Health Affairs* 28:5, w822–w831.
4   Ross A. Hammond and Ruth Levine. 2010. The economic impact of obesity in the United States. *Diabetes, Metabolic Syndrome and Obesity: Targets and Therapy* 3, 285–295.
5   Katherine Guthrie and Jan Sokolowsky. 2017. Obesity and household financial distress. *Critical Finance Review* 6:1, 133–178.
6   Gabriella Conti and James J. Heckman. 2010. Understanding the early origins of the education-health gradient: A framework that can also be applied to analyze gene-environment interactions. *Perspectives on Psychological Science* 5:5, 585–605.
7   John Cairney, Laurie Corna, Scott Veldhuizen, Paul Kurdyak, and David L. Streiner. 2008. The social epidemiology of affective and anxiety disorders in later life in Canada. *The Canadian Journal of Psychiatry* 53:2, 104–111.
8   Lenny R. Vartanian. 2010. Disgust and perceived control in attitudes toward obese people. *International Journal of Obesity* 34:8, 1302–1307.
9   Albert J. Stunkard, Myles S. Faith, and Kelly C. Allison. 2003. Depression and obesity. *Biological Psychiatry* 54:3, 330–337; Genevieve Gariepy, D. D. Nitka, and N. N. Schmitz. 2010. The association between obesity and anxiety disorders in the population: A systematic review and meta-analysis. *International Journal of Obesity* 34:3, 407–419.
10   Robert O. Bonow and Robert H. Eckel. 2003. Diet, obesity, and cardiovascular risk. *New England Journal of Medicine* 348:21, 2057–2133; Claude Bouchard, Jean-Pierre Depres, and Angelo Tremblay. 1993. Exercise and obesity. *Obesity Research* 1:2, 133–147.
11   Raffaella Molteni, Rachel Barnard, Zhe Ying, Christian K. Roberts, and Fernando Gomez-Pinilla. 2002. A high-fat, refined sugar diet reduces hippocampal brain-derived neurotrophic factor, neuronal plasticity, and learning. *Neuroscience* 112:4, 803–814.

12 M. Cristina Polidori, Domenico Praticó, Francesca Mangialasche, Elena Mariani, Olivier Aust, Timur Anlasik, Ni Mang Ludger Pientka, Wilhelm Stahl, Helmut Sies, Patrizia Mecocci, and Gereon Nelles. 2009. High fruit and vegetable intake is positively correlated with antioxidant status and cognitive performance in healthy subjects. *Journal of Alzheimer's Disease* 17:4, 921–927.

13 Charles H. Hillman, Kirk I. Erickson, and Arthur F. Kramer. 2008. Be smart, exercise your heart: Exercise effects on brain and cognition. *Nature Reviews Neuroscience* 9:1, 58–65; Arthur F. Kramer and Kirk I. Erickson, K. I. 2007. Capitalizing on cortical plasticity: Influence of physical activity on cognition and brain function. *Trends in Cognitive Sciences* 11:8, 342–348; Kirk I. Erickson, Michelle W. Voss, Ruchika Shaurya Prakash, Chandramallika Basak, Amanda Szabo, Laura Chaddock, Jennifer S. Kim, Susie Heo, Heloisa Alves, Siobhan M. White, Thomas R. Wojcicki, Emily Mailey, Victoria J. Vieira, Stephen A. Martin, Brandt D. Pence, Jeffrey A. Woods, Edward McAuley, and Arthur F. Kramer. 2011. Exercise training increases size of hippocampus and improves memory. *Proceedings of the National Academy of Sciences* 108:7, 3017–3022; Andrea M. Weinstein, Michelle W. Voss, Ruchika S. Prakash, Laura Chaddock, Amanda N. Szabo-Reed, Siobhan M. Phillips, Thomas R. Wójcicki, Emily L. Mailey, Edward Mcauley, Arthur F. Kramer, and Kirk I. Erickson. 2012. The association between aerobic fitness and executive function is mediated by prefrontal cortex volume. *Brain, Behavior, and Immunity* 26:5, 811–819.

14 Patrick J. Smith, James A. Blumenthal, Benson M. Hoffman, Harris Cooper, Timothy J. Strauman, Kathleen Welsh-Bohmer, Jeffrey Browndyke, and Andrew Sherwood. 2010. Aerobic exercise and neurocognitive performance: A meta-analytic review of randomized controlled trials. *Psychosomatic Medicine* 72:3, 239–252.

15 Shwan M. Arent, Daniel M. Landers, and Jennifer L. Etnier. 2000. The effects of exercise on mood in older adults: A meta-analytic review. *Journal of Aging and Physical Activity* 8:4, 407–430; Felice N. Jacka, Peter J. Kremer, Eva R. Leslie, Michael Berk, George C. Patton, John W. Toumbourou, and Joanne W. Williams. 2010. Associations between diet quality and depressed mood in adolescents: Results from the Australian healthy neighbourhoods study. *Australian and New Zealand Journal of Psychiatry* 44:5, 435–442.

16 H. S. Rosen and S. Wu (2004). Portfolio choice and health status. *Journal of Financial Economics* 72:3, 457–484.

17 Ryan D. Edwards. 2010. Optimal portfolio choice when utility depends on health. *International Journal of Economic Theory* 6:2, 205–225.

18 Motohiro Yogo. 2016. Portfolio choice in retirement: Health risk and the demand for annuities, housing, and risky assets. *Journal of Monetary Economics* 80:June, 17–34; Gaobo Pang and Mark Warshawsky. 2010. Optimizing the equity-bond-annuity portfolio in retirement: The impact of uncertain health expenses. *Insurance: Mathematics and Economics* 46:1, 198–209.

19 Dana Goldman and Nicole Maestas. 2013. Medical expenditure risk and household portfolio choice. *Journal of Applied Econometrics* 28:4, 527–550.

20 Silvia Bressan, Noemi Pace, and Loriana Pelizzon. 2014. Health status and portfolio choice: Is their relationship economically relevant? *International Review of Financial Analysis* 32:March, 109–122.

21 Jawad M. Addoum, George Korniotis, and Alok Kumar. 2017. Stature, obesity, and portfolio choice. *Management Science* 63:10, 3393–3413.

22 David A. Love and Paul A. Smith. 2010. Does health affect portfolio choice? *Health Economics* 19:12, 1441–1460.

23 Fernando Patterson and Corey Shank. 2019. Obesity, health habits, and financial decision making. Working Paper, Dalton State University.

24 John R. Nofsinger and Corey Shank. 2019. Community health and firm performance. Working Paper, University of Alaska.

25 Anup Agrawal and Yuree Lim. 2018. Local obesity prevalence and corporate policies. *Quarterly Journal of Finance* 8:2, 1–33.
26 Alex Edmans. 2011. Does the stock market fully value intangibles? Employee satisfaction and equity prices. *Journal of Financial Economics* 101:3, 621–640.
27 Alex Edmans. 2015. The social responsibility of business. Retrieved from www.youtube.com/watch?v=Z5KZhm19EO0, July
28 Peter Limbach and Florian Sonnenburg. 2015. Does CEO fitness matter? Working Paper, University of Cologne.
29 Brian C. McTier, Yiumen Tse, and John K. Wald. 2013. Do stock markets catch the flu? *Journal of Financial and Quantitative Analysis* 48:3, 979–1000.
30 Vicki L. Bogan and Angela R. Fertig. 2013. Portfolio choice and mental health. *Review of Finance* 17:3, 955–992.
31 Vicki L. Bogan and Angela R. Fertig. 2018. Mental health and retirement savings: Confounding issues with compounding interest. *Health Economics* 27:2, 404–425.
32 Maarten Lindeboom and Mariya Melnychuk. 2015. Mental health and asset choices. *Annals of Economics and Statistics* 119:120, 65–94.
33 Qiyan Ong, Walter Theseira, and Irene Y.H. Ng. 2019. Reducing debt improves psychological functioning and changes decision-making in the poor. *Proceedings of the National Academy of Sciences* 116:15, 7244–7249.
34 Angus Deaton. 2012. The financial crisis and the well-being of Americans. *Oxford Economic Papers* 64:1, 1–26.
35 Melissa McInerney, Jennifer M. Mellor, and Lauren Hersch Nicholas. 2013. Recession depression: Mental health effects of the 2008 stock market crash. *Journal of Health Economics* 32:6, 1090–1104.
36 Hannes Schwandt. 2018. Wealth shocks and health outcomes: Evidence from stock market fluctuations. *American Economic Journal: Applied Economics* 10:4, 349–77.
37 Anita Ratcliffe and Karl Taylor. 2015. Who cares about stock market booms and busts? Evidence from data on mental health. *Oxford Economic Papers* 67:3, 826–845.
38 Chad Cotti, Richard A. Dunn, and Nathan Tefft. 2015. The Dow is killing me: Risky health behaviors and the stock market. *Health Economics* 24:7, 803–821.
39 Corrado Giulietti, Mirco Tonin, and Michael Vlassopoulos. Forthcoming. When the market drives you crazy: Stock market returns and fatal car accidents. *Journal of Health Economics*.
40 Chad Cotti and David Simon. 2018. The impact of stock market fluctuations on the mental and physical well-being of children. *Economic Inquiry* 56:2, 1007–1027.
41 Joseph Engelberg and Christopher A. Parsons. 2016. Worrying about the stock market: Evidence from hospital admissions. *Journal of Finance* 71:3, 1227–1250.
42 Tomasz Piotr Wisiniewski and Brendan John Lambe. 2018. Do stock market fluctuations affect suicide rates? Working Paper, University of Leicester.
43 I. Kalcheva, P. McLemore, and Richard Sias. Forthcoming. Economic policy uncertainty and self-control: Evidence from unhealthy choices. *Journal of Financial and Quantitative Analysis*.
44 Howard Kahn and Cary L. Cooper. 1990. Mental health, job satisfaction, alcohol intake and occupational stress among dealers in financial markets. *Stress Medicine* 6:4, 285–298.
45 Howard Kahn, Cary L. Cooper, and Sharon P. Elsey. 1994. Financial dealers on Wall Street and in the city of London: Is the stress different? *Stress Medicine* 10:2, 93–100.
46 Thomas Oberlechner and Ashok Nimgade. 2005. Work stress and performance among financial traders. *Stress and Health: Journal of the International Society for the Investigation of Stress* 21:5, 285–293.

Section III
Cognitive Outcomes

9 The Emotional and Moody Investor

People make rational investment decisions because money is involved, or at least that is one of the pillars of traditional finance. However, research shows that the opposite is true—people make *irrational* investment decisions because money is involved. Two of the obstacles to making rational decisions may be emotions and mood. Emotions like anger, happiness, sadness, and empathy can have a powerful influence on your decisions and actions. People who are in a bad mood are more pessimistic about future events, including investment prospects. People who are in a good mood are optimistic about investment prospects. This chapter focuses on the impact of emotional responses and mood on financial behavior.

Emotions, Mood, and Physiology

Psychologists Paul Ekman and Wallace V. Friesen argue that we have six basic emotions: happiness, anger, sadness, fear, surprise, and disgust.[1] They state that all other emotions are just amalgamations of these six basic emotions, in the same way that all colors are simply a mix of three primary colors (red, yellow, and blue). For example, the emotion of amazement is the combination of surprise and happiness, while remorse is a combination of disgust and sadness. Anxiety is the combination of fear and sadness, while anger and sadness combine to trigger the feeling of betrayal.

Where does emotion originate? It is generated from a biological source—the limbic system of the brain. To biologically determine how an emotion might influence decision making, scientists first locate which parts of the brain are activated during emotional responses. Those decisions facilitated in that region are associated with the emotion. Phan and colleagues conducted a meta-analysis of emotion activation and neural activity by reviewing 55 different studies on the topic (See Chapter 5 about brain function for further details about these structures).[2] Following their review, they found strong evidence of the relationship between neural activity and emotion. The medial prefrontal cortex had a general role in all emotional processing. Happiness was related to activation in the ventral striatum and putamen and was related to the dopaminergic system, which suggests that happier people take more risk. Sadness was elicited by activation in the subcallosal cingulate, which is the part of the brain associated with depression, suggesting that sad people will exhibit avoidance tendencies and take less risk. Fear was strongly associated with increased activation of the amygdala, which triggers the body's fight or flight response. Depending on the person, fear can result in risk aversion (i.e., flight) or risk seeking (i.e., fight) behavior.

Emotions and Economic Decisions

Physiological Response to Economic Decisions

While emotions are often thought of as a cognitive reaction, they are also associated with a physical response. Indeed, emotions fall into a category of psychophysiology that combines the study of psychology and biology. To examine the psychophysiology of financial risk processing, Lo and Repin placed electrodes and other sensors onto professional traders during their working hours in order to measure their skin conductance, blood pressure, heart rate, respiration, and body temperature.[3] The authors found that investors physically reacted to changes in stock price changes and volatility through changes in skin conductance and body temperature. High volatility periods in the market caused increases in skin conductance and body temperature. Medical literature finds that body temperature increases when people experience negative emotions in response to stimuli. Even professional traders have emotional responses to changes in stock prices.

Emotions, Mood, and Experimental Finance

There is extensive literature on the impact of emotions on many facets of decision making. Psychologists and economists call these feelings *affect* or *emotional affective states*. Kuhnen and Knutson examined the impact of affect on financial decisions by manipulating emotional responses.[4] In this experiment, each person completed 90 investment decisions in which they selected between a risky security (stock) and a riskless security (bond). The stock either gained $10 or lost $10 while the bond paid out a guaranteed $3. After making each selection, the participants saw the outcome of their choice and were asked the following two questions: (1) "What do you think is the probability that the stock is the good stock?"; and (2) "How much do you trust your ability to come up with the correct probability estimate that the stock is good?" After each round, each participant was shown a picture that was designed to manipulate their emotional state. The three different pictures where designed to be highly arousing and positive (e.g., erotic scenes), highly arousing and negative (e.g., rotten food), and neutral (e.g., a book lying on the floor). When the participant was given a picture designed to create a negative affective state, they selected the riskier investment 36 percent of the time, which is significantly lower than when they were shown either the neutral image (41 percent) or the positive image (40 percent). Overall, these results suggest that emotional states may influence investors' risk taking.

Lerner and colleagues examined the impact of emotions on the endowment effect, which is the tendency to place a higher value on items that you own.[5] In this experiment, the subjects watched one of three short movie clips intended to manipulate their emotions (sadness, disgust, and neutral). The sadness clip is of a young boy's mentor dying (from the movie *The Champ*), the disgust clip is about a man using an unsanitary toilet (from *Trainspotting*), and the neutral clip is about a fish in the *Great Barrier Reef* (from a National Geographic special). After watching the clip, the subjects were split into two groups. The first group was given a set of highlighter pens and had to select between keeping the set or selling it for an amount of cash that ranged from $0.50 to $14.00. The second group was given a choice between taking various amounts of money or keeping the set of highlighter pens. The endowment effect argues that subjects place a higher value on items which they already own, so they believe that the set of highlighter pens will be worth more if they actually own it (i.e., group 1) compared to those who do not own it (i.e., group 2). Figure 9.1 presents a graphical representation of the authors' results. The neutral film clip elicited a typical yet expected result in that as the owners of an item we place a higher value on it than do people who want to buy that same item. However, the results from the disgust and sadness affects displayed the opposite result. People who owned the highlighter set put a lower value on it than those who wanted to buy it. One explanation for this is that after viewing the sad or disgusting movie clip, the participants associated those same feelings with the highlighter set (i.e., when they looked at the highlighter set they became sad or disgusted) and wanted to get rid of it so as to be free of those feelings.

In another study, Conte and colleagues examined the effect of the emotional states of joviality, sadness, fear, and anger on risk preferences.[6] Each

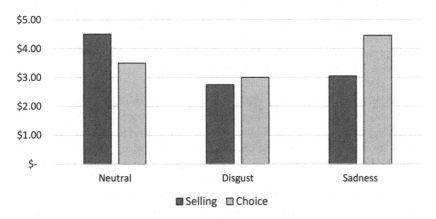

Figure 9.1 Affective Cues and the Endowment Effect

subject was randomly grouped into one of the four emotional states or into a neutral group. To elicit each emotion, the authors showed the subjects a clip of a film. To elicit joviality, they showed a clip from *When Harry Met Sally*, in which the two characters debate whether Harry would notice if a woman fakes an orgasm. To elicit sadness, the subjects watched a clip from the movie *The Champ*. A boxer is lying on a table gravely injured when his son enters the room and watches him die. Fear was elicited by a clip from *The Silence of the Lambs* in which a woman follows a killer into a dark basement. Anger was elicited by watching a man being attacked and beaten by a group of people. Finally, the neutral treatment was achieved by watching a clip of two men talking to each other in a courtroom in a scene from *All the President's Men*. After watching one of these clips, each subject selected between 100 pairwise lottery choices. Each of the choices had four possible outcomes—$0, $8, $16, and $24—with different probabilities. Overall, the authors found that people from all four affective states took more risk compared to the neutral group.

Lerner and Keltner examined the impact of fear and anger on risk taking.[7] To measure fear, the authors used a questionnaire, which asked questions about how the participant would react to specific situations such as being trapped in a small space or in a room with snakes. To measure anger, they provided a separate survey that asked how often subjects experienced anger and what was its intensity. The authors conducted two different experiments. In the first, participants were given a scenario that asked them to imagine that the United States was preparing for an Asian disease outbreak that was expected to kill 600 people and then decide if they believed program A or program B (gain frame) would be a better choice to combat this new disease. Other participants were provided the choice between program C and program D (loss frame). The programs were:

- If program A is adopted, 200 people will be saved.
- If program B is adopted, there is a 1/3 probability that 600 people will be saved and a 2/3 probability that no one will be saved.
- If program C is adopted, 400 people will die.
- If program D is adopted, there is a 1/3 probability that nobody will die and a 2/3 probability that 600 people will die.

Whether given a choice between A and B or between C and D, note that there is no correct answer. It is just a matter of preference. Programs A and C are riskless in that you know exactly what will happen. However, programs B and D involve risk because the outcome is determined by probability. Also, given the base case of 600 people dying, note that program A and program C both indicate that 400 people will die. The difference between A and C is not the number of deaths, it is the frame in which the outcome is offered. Program A uses a positive frame (saved lives), while program C uses a negative frame (deaths). Also note that programs B and D also provide the same outcome, but report it in different frames—positive versus negative. If these frames did not impact you, then you would pick the same program (A and C or B and D) that you had given as your preference no matter which frame you were presented with. For people in a neutral emotional state, the common result of this choice is that when presented with programs A and B, roughly three-fourths of the people select program A. In the frame of saving lives, people do not want to take the risk. In the negative frame choice, roughly three-fourths of the people choose program D over C. That is, most people choose the risky option when they are presented with people dying.

In addition to picking between programs A and B in the positive frame or between programs C and D in the negative frame, participants were asked about the conviction of their choice. Through a six-point Likert scale where 1 represents "very much prefer program A" and 6 represents "very much prefer program B," the authors found that the subjects who reported more fear in the survey were more likely to avoid risk and thus select program A. People who reported more anger were more likely to take more risk and select program B. Additionally, the authors found the same result when examining the negative frame as more fear (anger) resulted in participants being more likely to select program C (D). These results suggest that fear and anger are emotions that are not susceptible to the framing effect.

In the second experiment, the authors provided the same anger and fear surveys, but also gave subjects a happiness mood survey. The authors conducted the same experiment with these three different emotions; happiness, fear, and anger. The authors found the same results for anger and fear as before and the additional results that anger and happiness have nearly identical results. Both angry and happy participants were more likely to gamble with people's lives by selecting programs B and D.

Prospect theory predicts that people will become risk averse after winning money and risk seeking after losing money. This reaction is due to being happy when you win money and wanting to savor that happiness by not risking the gain. To avoid the sadness of losing money, people tend to increase the risk in order to get back to an even level. Additionally, the disposition effect is closely tied to prospect theory, which is the behavior of investors selling their winning positions too soon (i.e., being risk averse with gains) and holding onto their losers for too long (i.e., being risk seeking with losses). Summers and Duxbury examined the impact of emotions on the disposition effect.[8] In their experiment, each subject was given ten shares of a stock and watched the stock go up or down over one period. After this one period, they had the option of selling, holding, or buying shares. After making the decision to buy, hold, or sell, the participants were asked how satisfied they were with their decision. The authors found that people were happier after selling for a profit and disappointed after selling for a loss. Additionally, they found that the disposition effect only occurred when participants were emotional, which suggests that the disposition effect is closely tied to emotion and mood.

Seo and colleagues examined the influence of affect on investment risk and investment decision frames.[9] They used an investment simulation whereby there was a session every day for 20 consecutive business days. For each day's simulation, the participants undertook three tasks: (1) they viewed market and stock information for 12 anonymous stocks; (2) they checked their current portfolio performance; and (3) they made decisions about buying or selling stocks for that day. In between the tasks, the subjects were asked to rate their feelings. Participants' level of pleasant feelings could be rated as low (sad), normal, or happy. These affective states were then examined in relation to the realization of prior trades and new trades made for the day. Without taking into consideration affect, behavioral finance shows that people tend to take less risk after experiencing an investment gain. On the other hand, people tend to take more risk after an investment loss. Thus, there is a framing effect on the reaction to gains and losses. How does affect impact risk taking and the framing effect? The authors examined this question by creating a measure of risk that relied on both the riskiness of the individual securities owned and the degree of diversification. The measure was then scaled in reference to the risk of the market in general so that a risk of zero was the same as the market portfolio. We index their measure and report it in Figure 9.2 for two cases of the gain frame and two cases of the loss frame. It is useful to examine the reaction of participants experiencing normal pleasantness. Note that they took less risk after gains and more risk after losses, as expected. However, the authors found that affective states can mitigate this result. Specifically, after a gain, happy participants did not take less risk. Indeed, they took marginally more risk. After a loss, happy people did take risk, but it was less risk than people with normal pleasantness. Thus, the authors concluded that affective state did not impact risk taking in general because happiness caused more risk in one frame and less risk in another. However, affective state did mitigate the framing effect, while sadness exacerbated it.

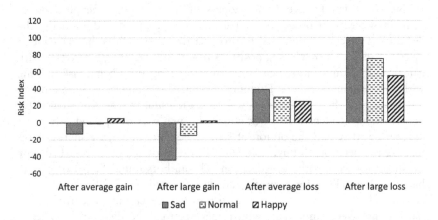

Figure 9.2 Affective States and Risk Taking in Gain/Loss Decision Frames

Andrade and colleagues examined the impact of emotion on stock market bubbles.[10] In this experiment, the participants had the opportunity to buy and sell stock through 15 rounds of trading. At the end of each round, they received a dividend for 0, 8, 28, or 60 cents per share. The expected dividend was 24 cents (the average of 0, 8, 28, and 60) which resulted in the stock having a fundamental value of $3.60 (24 cents × 15 rounds). Additionally, prior to the trading experiment each participant watched a short video clip. To elicit excitement, participants watched action scenes from the movies *Knight and Day* and from *Mr. and Mrs. Smith*. To elicit calmness, the participants watched a clip from *Franklin and Peach in the Water*. Finally, to elicit fear, participants watched video clips from *Hostel* and *Salem's Lot*. The authors found that excited investors were much more likely to create bigger bubbles (i.e., they pay more for the stock than it is worth) and had longer lasting bubbles (i.e., they stayed at a higher price for longer).

Loss Aversion and Affect

Myopic loss aversion depends on two behavioral biases: loss aversion and mental accounting. Loss aversion refers to the fact that people are more sensitive to losses than they are to gains (i.e., losing $50 feels worse in magnitude than winning $50 feels good). Indeed, people behave in seemly irrational ways in order to avoid taking a loss. Mental accounting is the cognitive bias on which people tend to consider each investment individually rather than all together as a portfolio. Investors examine the gain/loss position of each stock, mutual fund, and retirement account separately. Putting the two together, *myopic loss aversion* is simply the motivation from these two tendencies to avoid losses in each individual investment account.

Thaler and colleagues examine the impact of myopia and loss aversion on risk taking.[11] The authors conducted an experiment in which participants must allocate 100 shares between fund A and fund B and then get paid according to the success of their portfolio decisions. Fund A had an expected return of 0.25 percent with a standard deviation of 0.177 percent, so it was not possible for its return to be negative. Fund A simulated a typical bond mutual fund. Fund B had an expected return of 1 percent and a standard deviation of 3.54 percent, which corresponds to a typical stock mutual fund. Subjects were randomly selected into either the monthly, yearly, five-yearly, or inflation monthly groups. In the monthly group, the subjects made buy, sell, or hold decisions and then watched their portfolio simulation over one period. They traded again and viewed the outcome, repeating this process for a total of 200 trading decisions. In the yearly group, the subjects traded and watched their portfolio simulation over eight periods, repeating for a total of 25 trades and outcomes. The five-yearly group traded and watched their portfolio simulation for 40 periods repeating for a total of five times. Finally, the inflation monthly group experienced the same experiment as the monthly group except that their returns were translated upward by 10 percent, so these subjects always had a positive return for both funds.

The idea behind myopic loss aversion is that the less frequently investors check and rebalance their portfolio, the less emotional they will be when they trade. Given that the stock fund has an average monthly return of 1 percent, the chances of having negative returns over a long period of time (i.e., years) is very low and roughly 0.3 percent. However, the probability of any single month be negative is roughly 39 percent. Therefore, if investors are more likely to take less risk following losses due to possible sadness, they are more likely to invest in less risky assets if they check their portfolios more often. Figure 9.3 depicts the average amount the subjects invested in fund A (the bond fund) after learning through the many trades and outcomes. Specifically, it shows the allocation to bonds during the final five years of simulation and just their final decision for each group. These results show that the participants who checked and traded the most frequently (i.e., monthly) decided to take less risk than the other participants by investing more in the bond fund (fund A).

Haigh and List conducted an experiment in which they examined myopic loss aversion in 32 students and 27 professional stock investors.[12] Due to training and experience, professionals should be able to control their emotions better than students who are investment novices. Both groups participated in two treatments whereby they were given $100 in each round and asked how much (or all of it) they wanted to gamble on a bet that had a 1/3 probability of earning 250 percent and a 2/3 probability of losing the amount offered. In treatment 1, the subjects made decisions throughout nine consecutive rounds, which was called the frequent decision treatment. In treatment 2, the subjects made infrequent reallocation decisions after only rounds three and six. The principle of myopic loss aversion dictates that the subjects will bet lower amounts the more frequently they make decisions as they will become more emotional with more frequent feedback. Figure 9.4

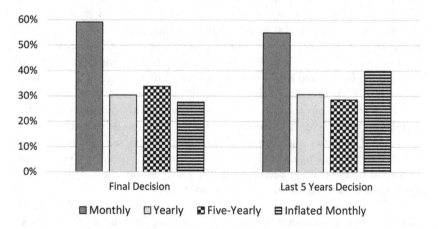

Figure 9.3 Rebalancing Frequency and Bond Fund Allocation

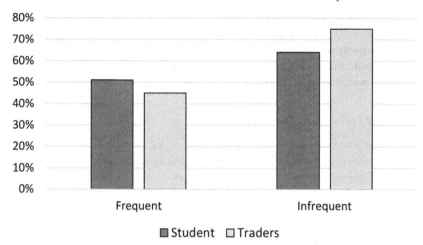

Figure 9.4 Average Amount Bet Based on Treatment and Experience

presents a graphical representation of the study results. First, the authors found that more frequent feedback resulted in lower bets, consistent with myopic loss aversion. However, rather than the result most people would expect, the authors found that professional traders exhibited more myopic loss aversion than did novice student investors.

Sokol-Hessner and colleagues examined whether people can purposely regulate their emotions during investment decisions.[13] Specifically, they studied the impact of emotional regulation on loss aversion. Emotional activation was measured in two ways. First, emotions elicit physiological responses that can be measured through skin conductance, a measure of sympathetic nervous system activity. Skin conductance was recorded during the choice task as a measure of arousal. The second measure was through behavioral outcome. Participants experiencing higher emotional arousal exhibit greater loss aversion. Thus, the extent of loss aversion during the experiment was measured. In this experiment, the subjects were given $30 and had to make gambles across 280 trials. Of their decisions, 240 had a random chance of earning money or losing money, while 40 of their decisions had a guaranteed gain. Prior to making a selection, the subject was given one of two cognitive regulation strategies. The "attend" strategy instructed the subjects to consider each gamble in isolation from all other decisions as if each decision was the only one they would make (mental accounting) and to let any emotions occur naturally. The "regulate" strategy instructed the subjects to consider each decision in the context of the bigger picture that there are 280 trials and to think like a professional trader and to not make emotional decisions. The study had three critical conclusions:

- Participants who "thought like a trader" exhibited lower emotional responses as physiologically measured through skin conductivity.
- Participants who "thought like a trader" also exhibited lower loss aversion.
- The more that participants were able to regulate their emotional response as measured by skin conductivity, the lower their loss aversion behavior.

Affect and Ethical Decision Making

Given that emotions influence risk taking and rational thinking, it is also likely that emotions will influence ethical decision making. Coricelli and colleagues examined this idea in the context of tax evasion.[14] They conducted an experiment whereby subjects were randomly given either $50, $100, $150, or $200. Following the random endowment, the subjects were required to report their income, which would be taxed at a rate of 55 percent. The subjects knew that there was a chance that they might be audited. The audit would require people who did not report their income to pay a fine higher than the tax rate. If the subject reported a high income, there was a 65 percent chance that they would be audited. If they reported a low income, there was a 35 percent chance that they would be audited. Thus, if all the subjects reported the same income, there was a 50 percent chance of their being audited. If the subject was not audited, or was audited and found to have correctly reported, they kept their income minus taxes. If the subject was audited and found to be guilty of tax evasion, they paid a 20 percent penalty in addition to the correct amount of taxes owed. Additionally, the subjects were given a survey to assess their emotional state and a skin conductance was carried out during the session to measure physiological changes. The study found that subjects who showed increased skin conductance response, which indicates a higher anticipatory emotional arousal, were more likely to commit tax evasion. This result suggests that people guilty of tax evasion display signs of fear and anxiety due to the moral implications of their deception.

Maciejovsky and colleagues examined the impact of emotions and ethics through cash-based businesses.[15] Businesses that rely heavily on cash transactions are often susceptible to having lower ethical standards. In this experiment, the subjects were put through an emotional priming survey. The survey asked them to do simple math to get them thinking about rational decisions. The emotional portion of the survey asked questions about how they felt about certain things, such as politics. Following this, they were asked how likely they would be to complete a sale without issuing an invoice. The studies found that those people who were more emotional were more likely to not issue an invoice and not be tax compliant.

Aggregate Mood and the Stock Market

One criticism of the previous experimental research is that it is hard to forecast how these individual differences at the micro level can impact the macro level, such as the stock market or economy as a whole. As such, it has been important for researchers to find ways to aggregate the mood of individuals in order to calculate the mood of the general population.

Da and colleagues proxy the investor mood of the general society by using internet search volume as it relates to the stock market.[16] They categorized words from the Google search engine and placed them into categories such as "positive," "negative," "weak," and "strong." Furthermore, as they were concerned only with the stock market mood, they selected words that were related to the stock market or economy. In total, they used 149 search words, such as "bankruptcy," "crisis," "gold," "inflation," "depression," or "recession." After compiling the list of words that related to investor mood, they determined the search volume trends of these words during various aspects of economic and stock market performance. More searches for terms such as "recession" or "depression" indicated that more investors were worried about poor stock returns and that the investor mood was poor. The authors found that this search index can predict short-term stock return reversals and increases in volatility and fund flows in and out of equity and bond funds. Overall, these results show that aggregate investor mood can influence stock price and market behavior. That is, people react to their moods.

If the general mood of society can impact the stock market, then some people might try to influence that mood. For example, company executives might try to create a positive mood surrounding their firm. Opportune times would be during organized company conference calls and earnings announcements. Some of the most active days for stock trading are around company quarterly earnings announcements. The announcements are scheduled in advance, and many investors pay attention. During these announcements, the company will release essential numbers such as revenue for the quarter, earnings per share, and some expectations about the upcoming quarter. The firm may try to "sugarcoat" the numbers and generally create a rosy ambiance. Does this effort succeed in creating positive sentiment? To better understand this dynamic, research has examined what the chief executive officer (CEO) has to say and how it is said.

Price and colleagues examined the transcripts of various earnings conference calls to determine the sentiment of the CEO.[17] To do this, they counted the number of positive and negative words. They created a word-based sentiment index to determine if the CEO was more or less optimistic during their speech. The study found that the more positive words the CEO used, the higher the stock returns the next trading day. Thus, CEO sentiment can impact short-term stock prices.

Does this sentiment influence all investors? Blau and colleagues expanded the prior study by examining the number of positive and negative words from the conference calls and the reaction of sophisticated investors versus normal investors.[18] The authors argued that sophisticated investors, as proxied by professional investors and analysts, had more training and experience to interpret the information provided. To differentiate between sophisticated and average investors, they examined short sales. Short selling is a complicated trade in which the trader benefits from the decline of a stock price. Because of the risks involved, only advanced investors use short selling. The authors found that short sellers target firms with greater earnings surprise (higher earnings than expected) and an unusually high amount of positive words used in the CEO's speech. In other words, sophisticated investors recognize the CEO's attempt to inflate the firm's stock price through sentiment. Short sellers try to take advantage of a temporary short-term price increase and then sell short to profit from the price reversion.

In another study, Price and colleagues analyzed audio recordings and the corresponding written transcripts of the CEO.[19] They used Audacity software and a Layered Voice Analysis software platform to isolate vocal characteristics, such as if the CEO was excited and happy during the speech or depressed and unhappy. The authors found that the emotion of the CEO was positively related to how investors viewed the information. That is, if the CEO was happy and upbeat during the conference call, stock returns were generally higher. This result suggests that it is not only what the CEO says, but also how they say it that matters.

Mayew and Venkatachalam examined affective states during conference calls and future firm performance.[20] The authors used Layered Voice Analysis software to measure positive and negative affective (emotional) states. This software used the tone of the executives' voices to determine if a positive affective state (i.e., happy, excited, joyful) or a negative affective state (i.e., fearful, tense, anxious, unhappy) was displayed. However, it was possible that the CEO could fake a positive affective state while reading a speech during the conference call. Therefore, the authors examined the affective state of the CEO or chief financial officer during their response to analysts' questions. They argued that affective states will be most apparent during the question and answer portion of the conference call. These authors found that investors reacted immediately to negative affect after the company missed earnings as stock prices dropped. Additionally, they found that when analysts later changed their stock recommendations, they incorporated positive rather than negative affect.

Summary

The sufficiently rational and emotionless "homo economicus" does not exist. Research on emotion and mood demonstrates that emotions influence people's propensity to take on more or less risk. For example, people who are happier or sadder are more likely to take more risk. Additionally, more frequent checking of one's portfolio and investment positions fosters increased emotional arousal to losses, which influences individuals to take more risk in the future. This emotional response to gains and losses is also shown to influence the disposition effect, as those who are less emotional are less likely to commit the investment bias. Furthermore, the mood of society has an impact on stock market dynamics. For instance, when people are worried about the stock market and are searching Google for negative words, the stock market tends to drop. Conversely, when the CEO is in a positive mood during company earnings announcements the stock price tends to increase afterwards. Overall, research indicates that mood and emotion can have a large impact on financial decision making and stock market movements.

Questions

1 How do emotions effect risk taking?
2 How do emotions effect irrational behavior?
3 What is myopic loss aversion and how does it relate to emotions?
4 How do researchers examine the mood of the aggregate stock market?

Notes

1 Paul Ekman and Wallace V. Friesen. 1971. Constants across cultures in the face and emotion. *Journal of Personality and Social Psychology* 17:2, 124.
2 K. Luan Phan, Tor Wager, Stephan F. Taylor, and Israel Liberzon. 2002. Functional neuroanatomy of emotion: A meta-analysis of emotion activation studies in PET and fMRI. *Neuroimage* 16:2, 331–348.
3 Andrew W. Lo and Dmitry V. Repin. 2002. The psychophysiology of real-time financial risk processing. *Journal of Cognitive Neuroscience* 14:3, 323–339.
4 Camelia M. Kuhnen and Brian Knutson. 2011. The influence of affect on beliefs, preferences, and financial decisions. *Journal of Financial and Quantitative Analysis* 46:3, 605–626.
5 Jennifer S. Lerner, Deborah A. Small, and George Loewenstein. 2004. Heart strings and purse strings: Carryover effects of emotions on economic decisions. *Psychological Science* 15:5, 337–341.
6 Anna Conte, M. Vittoria Levati, and Chiara Nardi. 2018. Risk preferences and the role of emotions. *Economica* 85:338, 305–328.
7 J. S. Lerner and D. Keltner. 2001. Fear, anger, and risk. *Journal of Personality and Social Psychology* 81:1, 146.
8 B. Summers and D. Duxbury. 2012. Decision-dependent emotions and behavioral anomalies. *Organizational Behavior and Human Decision Processes* 118:2, 226–238.
9 M.G. Seo, B. Goldfarb, and L.F. Barrett. 2010. Affect and the framing effect within individuals over time: Risk taking in a dynamic investment simulation. *Academy of Management Journal* 53:2, 411–431.
10 E. B. Andrade, T. Odean, and S. Lin. 2015. Bubbling with excitement: An experiment. *Review of Finance* 20:2, 447–466.
11 Richard H. Thaler, Amos Tversky, Daniel Kahneman, and Alan Schwartz. 1997. The effect of myopia and loss aversion on risk taking: An experimental test. *The Quarterly Journal of Economics* 112:2, 647–661.
12 Michael S. Haigh and John A. List. 2005. Do professional traders exhibit myopic loss aversion? An experimental analysis. *Journal of Finance* 60:1, 523–534.
13 Peter Sokol-Hessner, Ming Hsu, Nina G. Curley, Mauricio R. Delgado, Colin F. Camerer, and Elizabeth A. Phelps. 2009. Thinking like a trader selectively reduces individuals' loss aversion. *Proceedings of the National Academy of Sciences* 106:13, 5035–5040.

14 Giorgio Coricelli, Mateus Joffily, Claude Montmarquette, and Marie Claire Villeval. 2010. Cheating, emotions, and rationality: An experiment on tax evasion. *Experimental Economics* 13:2, 226–247.

15 Boris Maciejovsky, Herbert Schwarzenberger, and Erich Kirchler. 2012. Rationality versus emotions: The case of tax ethics and compliance. *Journal of Business Ethics* 109:3, 339–350.

16 Zhi Da, Joseph Engelberg, and Pengjie Gao. 2014. The sum of all FEARS investor sentiment and asset prices. *Review of Financial Studies* 28:1, 1–32.

17 S. McKay Price, James S. Doran, David R. Peterson, and Barbara A. Bliss. 2012. Earnings conference calls and stock returns: The incremental informativeness of textual tone. *Journal of Banking & Finance* 36:4, 992–1011.

18 Benjamin M. Blau, Jared R. DeLisle, and S. McKay Price. 2015. Do sophisticated investors interpret earnings conference call tone differently than investors at large? Evidence from short sales. *Journal of Corporate Finance* 31:April, 203–219.

19 S. McKay Price, Michael J. Seiler, and Jiancheng Shen. 2017. Do investors infer vocal cues from CEOs during quarterly REIT conference calls? *The Journal of Real Estate Finance and Economics* 54:4, 515–557.

20 William J. Mayew and Venkatachalam. 2012. The power of voice: Managerial affective states and future firm performance. *Journal of Finance* 67:1, 1–43.

10 How Environmental Factors Impact Financial Decisions

Most people go through life without realizing that environmental factors can impact their decisions. Have you ever had the rainy day blues and felt less motivated to work? Have your allergies ever been so bad that you didn't want to do anything except stay in bed? This chapter will detail how environmental impacts such as weather, pollution, and natural disasters influence financial decisions.

Weather, Mood, and Financial Decisions

Weather and Mood

Psychologists have found a robust correlation between the weather and mood. For example, most people are happier and more optimistic when there is sunshine rather than rain and clouds. Specifically, Howarth and Hoffman collected mood data on people over 11 consecutive days and matched it with weather data.[1] The authors found that on days with better weather, such as higher temperatures and hours of sunshine, the participants were in better moods. Hannak and colleagues even found that weather affects the sentiment of people's Twitter posts.[2] Additionally, increased amounts of sunshine are correlated with more helping behavior,[3] higher restaurant tipping,[4] and lower levels of depression[5] and suicide.[6] Thus, it would be expected that the weather would influence financial decisions through its impact on mood (see Chapter 9 on emotions and mood).

What causes this behavior? Sunshine causes the brain to increase the production of a neurotransmitter called serotonin. Serotonin relays signals between neurons in the brain that regulate happiness, anxiety, and other facets of mood, as well as bowel function, blood clotting, bone density, and sexual function. Low levels of serotonin, which may be caused by lack of sunshine among many other causes, leads to reduced happiness, higher anxiety, depression, and difficulty sleeping (See Chapter 7 on sleep). Conversely, darkness or darker lighting causes the body to produce a hormone called melatonin which is produced by the pineal gland. Melatonin is often called the sleep hormone as it influences your circadian rhythm. A rise in melatonin levels tells the body that it is time to sleep. Furthermore, melatonin binds to brain receptors to reduce nerve activity and dopamine levels that are trying to keep the body alert and awake.

Weather and Experimental Finance

Bassi and colleagues examined the impact of weather on financial risk taking through a lottery gamble task known as the Holt and Laury task.[7] During this task, participants were given ten possible decisions whereby option A was a chance to win $2 or $1.60 with varying probabilities and option B was a chance to win $3.85 or $0.10 with varying probabilities. The participants took this same task on two different days. One of the experiments took place during sunny weather while the other was during a rainstorm. The study found that on poor weather days, the participants were much more likely to select the less risky option (A) than when it was nice and sunny outside.

Weather and Stock Markets

Hirshleifer and Shumway examined the influence of weather and stock returns.[8] They argued that if good weather makes people more optimistic, then investors would be more likely to buy a stock than sell it, thus causing stock prices to increase. To test this hypothesis, they looked at the relationship between the amount of morning sunshine at the city headquarters of the domestic stock exchange in 26 countries, such as New York City in the United States. They found that sunshine had a positive effect on stock returns in 25 of the 26 countries (Manila in the Philippines was the exception).

Goetzmann and colleagues examined the impact of weather on investor perceptions of the stock market.[9] They used the Investor Behavioral Project Survey from Yale University that surveys approximately 100 professional traders per month. The survey asked investors about their perception of the stock market. The study matched these surveys with the weather on the day the investor completed the survey. The authors reported that on more cloudy days (i.e., less sunshine) investors perceived stocks as being more overpriced and found that professional trading institutions were more likely to sell stocks.

Seasonal Affective Disorder

Seasonal Affective Disorder and Physiology

Seasonal affective disorder (SAD) is a syndrome described as recurrent bouts of depression that occur due to there being less sunlight during daytime (September/October to April/May in the northern hemisphere). It affects about 10 percent of the population in the United States. However, it is possible that it can have a milder influence than actual clinical depression on a more significant portion of the population. This form of SAD is commonly referred to as the "winter blues."

SAD appears to have a neurological impact that changes behavior. For example, the medical literature suggests that the two leading causes are from lower levels of serotonin due to less exposure to daylight[10] and a disruption in the body's internal clock, often referred to as the circadian rhythm, because the time of the day the sun rises and sets changes.[11] Additionally, the change in circadian rhythm leads to increased sleep latency, which is the amount of time it takes to fall asleep, and less short-wave (delta) sleep often referred to as deep sleep.[12]

Cohen and colleagues examined the impact of seasonal affective disorder on neural activity using positron emission tomography (PET) scans.[13] The authors explained that people suffering from SAD may display different behaviors due to their having a lower glucose metabolic rate in their gray matter and in the prefrontal cortex, the parts of the brain that processes information and decision making. In such cases the brain cells do not convert the glucose in the cells to make energy. This makes cognitive processing slower and less efficient. Molin and colleagues examined the effect of weather on patients with SAD by surveying their mood every two weeks from September to May.[14] They found that in addition to the amount of daylight, the amount of sunshine and the temperature also played a significant role in mood and emotions. Thus, any study examining the impact of SAD must also account for other weather variables.

Seasonal Affective Disorder and Experimental Finance

Kramer and Weber conducted an experiment using university faculty and staff to assess the impact of SAD on financial risk aversion.[15] The participants who took part in the task were endowed with $20. They were then given a choice to gamble either 100 percent, 50 percent, or 0 percent of their endowment in a gamble that gave them a 50 percent probability to earn more than their gamble and a 50 percent probability to earn less than their gamble, but the expected outcome would be slightly greater than the amount of their gamble. The participants took part in this task in December 2008 and July 2009 to examine the influence of the number of daylight hours on risk aversion. The authors found that people who displayed more symptoms of SAD were less likely to gamble during the winter compared to the summer, suggesting that the number of daylight hours can influence risk taking.

Seasonal Affective Disorder and Stock Markets

Kaplanski and colleagues examined the impact of SAD and investment expectations.[16] They used the Longitudinal Internet Studies for the Social Sciences (LISS) panel data from Tilburg University to examine stock portfolios and investment expectations of adults in the Netherlands. The LISS uses a questionnaire that asks investors about their expectations for future stock returns and volatility (risk) in both the short term (next month) and the long term (next year) on the Amsterdam Exchange Index, which is the Dutch equivalent to the Dow Jones Industrial Average in the United States. The authors found that investors had lower stock return expectations during the fall months than they did throughout the rest of the year, suggesting that SAD reduced investors' mood and caused pessimistic views about the future. Similarly, Kamstra and colleagues examined the impact of SAD and stock market returns in the United States, Sweden, the United Kingdom, Germany, Canada, New Zealand, Japan, Australia, and South Africa using the amount of daylight in the location of the stock exchange as their measure of SAD.[17] The authors found that the number of hours of sunlight was inversely related to stock returns and showed that stock returns were lower in the fall and winter (in the northern hemisphere) than during the rest of the year.

If SAD is related to stock market returns, it may impact the return generating process. The Capital Asset Pricing Model is a mathematical model of portfolio theory and equilibrium in financial markets. It was independently developed by William Sharpe, John Lintner, Jack Treynor, and Jan Mossin. The model argues that expected equity returns are equal to the risk-free rate plus a risk factor (beta) multiplied by the market risk premium (stock market return minus risk free rate) of the stock market. The study won William Sharpe the Nobel Prize for Economic Sciences in 1990. Since the development of this model in the 1960s, researchers have attempted to improve upon it. For example, research has found that the market risk premium is not constant across time. This led to the development of the Conditional Capital Asset Pricing Model, which argues that the market risk premium changes over time. For example, the risk premium may be different in economic recessions compared to expansions.

Garret and colleagues examined the influence of SAD in explaining this time-varying market risk premium in the United States, the United Kingdom, Japan, New Zealand, Australia, and Sweden.[18] The study calculated the market risk premium for these countries using both daily and monthly price data and ran statistical regressions using the risk premium as the independent variable. The authors found that the amount of sunlight was positively correlated with the market risk premium suggesting that (1) SAD can influence the market risk premium; and (2) investors require lower returns in the fall and winter (in the northern hemisphere).

Another way to examine the risk of the equity market is to look at the bid-ask spread. The bid-ask spread refers to the difference in price paid by the buyer when buying shares of stock (the ask) and the price received by the seller (the bid). The ask price is higher than the bid price, representing one of the biggest transaction costs in the equity market. In 2019, the average spread was about 50 basis points for stocks trading on the New York Stock Exchange and 250 basis points for stocks listed on the Nasdaq. The bid-ask spread is highest in the trading of smaller firms.[19] Research shows that this spread changes throughout the year.[20] Additionally, risk aversion is a determinant in the bid-ask spread.[21] DeGennaro and colleagues examined whether SAD played a role in variations of the bid-ask spread, given its time-varying nature and relationship to risk aversion. The authors examined the bid-ask spread of all stocks trading on the New York Stock Exchange and Nasdaq markets and its relationship to the hours of sunlight in New York City (the location of the headquarters of both exchanges). The authors found that the bid-ask spread was 20 basis points wider during the fall and winter months, consistent with the idea that SAD increased risk aversion.[22]

In addition to examining risk aversion in bid-ask spread behavior, scholars can study it through investor behavior. Kamstra and colleagues examined investors' risk aversion by looking at the flow of money into and out of mutual funds.[23] The authors found that investors, in aggregate, move money into safer fund categories, such as money markets, during the fall and then back into riskier categories, such as equities, in the spring. The study argued that this finding was consistent with SAD as investors became more risk averse during the fall when the amount of daylight decreased and more risk seeking during the spring when the number of daylight hours increased.

In addition to studying the risk premium and risk aversion, scholars have directly examined risk through the market's return volatility. The higher the volatility, the higher the risk. Symeonidis and colleagues examined the influence of weather, including temperature, precipitation, sunshine, and hours of daylight on stock market volatility in 26 different countries.[24] They found that sunshine and the amount of daylight were positively related to stock market volatility in all 26 countries.

Seasonal Affective Disorder and Financial Analysts

Dolvin and colleagues examined whether there was a sunlight-based pattern in financial analysts' behavior that is similar to what we see with investors.[25] The authors examined the difference between a stock analyst's one-year forecast of earnings per share and the company's actual earnings per share. Additionally, the study matched these earnings forecasts to the amount of sunlight when the forecast was made. The authors found that financial analysts were more pessimistic and underestimated future earnings when there was less daylight and were more optimistic and overestimated their forecasts when there was more daylight. This evidence suggests that seasonal affective disorder plays a role in the accuracy of financial analysts' forecasts. Lo and Wu also examined the influence of SAD on financial analysts.[26] They found that SAD influenced analysts' annual earnings forecasts and reported earnings revisions. Revisions in the fall were more pessimistic, and revisions in the spring were more optimistic.

Dolvin and Pyles examined the influence of SAD on the pricing of initial public offerings (IPOs).[27] The authors hypothesized that if SAD caused increased risk aversion there would be higher underpricing when there was less daylight. Additionally, they hypothesized that due to the same risk aversion, offer price revisions and price adjustments would be larger in the fall and winter. Consistent with these hypotheses, the authors found that underpricing and price revisions were higher in months with fewer sunlight hours in a similar way to SAD influencing financial analysts' decisions on IPOs.

Air Pollution, Cognition, and Financial Decisions

Air Pollution and Cognition

The medical literature demonstrates that air pollution has a negative impact on cognitive abilities. However, there are multiple forms of air pollution, including sulfur dioxide, nitrogen dioxide, carbon monoxide, ozone, and particulate matter, also known as particulate pollution. Sulfur dioxide is a gas that can cause breathing irritation and difficulty, especially in people with pre-existing issues such as asthma. Nitrogen dioxide is a gas that can irritate the respiratory system and is primarily produced by the burning of fuel (i.e., cars, trucks, airplanes). Carbon monoxide is an odorless gas that forms when the fuel in carbon does not completely burn. Vehicle exhaust contributes approximately 75 percent of all carbon monoxide emissions. Ozone is a gas, but unlike the previously listed gasses, this one can be either good or bad. Ozone is naturally present more than six miles above the surface of earth and helps to shield the planet from the ultraviolet rays of the sun. However, ozone also forms near the ground from emissions from sources such as vehicles, power and chemical plants, and other industrial processes. Particulate matter consists of solid particles and liquid droplets in the air caused by dust, dirt, or smoke. Particulate matter can be grouped into PM_{10} or $PM_{2.5}$, which describes the approximate diameter of the pollutant. Specifically, particulate matter 10 refers to particulate matter less than 10 micrometers in dimension, while particulate matter 2.5 refers to particulate matter 2.5 micrometers or smaller in dimension. How small is 10 or 2.5 micrometers? The average human hair is 70 micrometers in diameter making single particle matter invisible to the naked eye. The United States Environmental Protection Agency (EPA) distinguishes between particulate matter 10 and 2.5. The finer the pollutant (i.e., $PM_{2.5}$), the more easily it can be drawn deeper into the lungs and cause more significant problems. The amount of particulate matter in the air is measured in micrograms per cubic meter ($\mu g/m^3$). The EPA uses these five pollutants (sulfur dioxide, nitrogen dioxide, carbon monoxide, ozone, and particulate matter) to create an air pollution index.

As air pollution passes through the respiratory system it inflames the neurons and creates oxidative stress, which is the imbalance between free radicals and antioxidants in the body.[28] This inflammation and oxidative stress, in turn, can change the function of mitochondrial DNA leading to damaged cells that replicate too slowly or incorrectly due to the lack of energy production from the mitochondria. In turn, this can cause brain damage and diseases such as Parkinson's or Alzheimer's, pulmonary and cardiovascular diseases, or many different forms of cancer.[29] Additionally, the impairment of air pollution reduces life expectancy[30] and increases illnesses and hospitalizations.[31] Overall, air pollution has a significant effect on the human body and brain function.

Zhang and colleagues examined the impact of the air pollution index on cognitive performance in Chinese people.[32] The study uses a longitudinal survey known as the China Family Panel Studies and asks 24 math questions and 34 word recognition (verbal) questions. The authors matched the math and verbal scores of the people to the air quality index of their location of residence. They found that air pollution was associated with lower cognition as measured by both lower scores on the math and verbal tests. Does air pollution have a long-term impact on cognition? Chen examined the impact of air pollution on cognitive development in China.[33] The author recorded the air pollution that students endured throughout their adolescence to see if it had an impact on their academic performance in later years. The study reported that children who grew up in areas with greater air pollution had lower reading ability and math skills as they developed, demonstrating long-term effects of air pollution on cognitive abilities.

However, outside pollution is not the only concern. Roth examined the effect of indoor air pollution on cognitive performance in the United Kingdom.[34] To do this, the author collected readings of particulate matter (10 and 2.5) and matched it with university students' exam scores during the 2012–2013 academic year. Roth found that students performed worse on their exams when pollution inside was higher than normal.

Air Pollution and the Stock Market

Levy and Yagil used the air pollution index from the EPA and split daily air pollution into good air quality and unhealthy air quality. They matched the good or unhealthy air quality of New York to the S&P 500, Dow Jones, and Nasdaq stock indices as all three are headquartered in the city. Figure 10.1 charts the average stock returns on both good air and unhealthy pollution

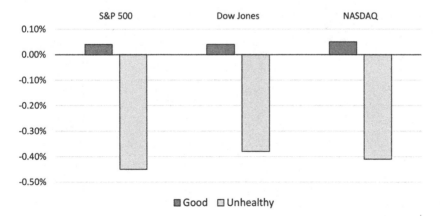

Figure 10.1 Daily Stock Return and Air Quality

days for each of the stock markets. The three different stock indexes aver-
aged about 0.05 percent per day when the air quality in New York was good
and between –0.38 percent and –0.45 percent on days when the air quality
was unhealthy.[35]

Heyes and colleagues used a more sophisticated model to examine the
influences of fine particulate matter ($PM_{2.5}$) on S&P 500 returns and volatility.[36] The average $PM_{2.5}$ in the Manhattan air was 11.53 $\mu g/m^3$. The study
reported that when particulate matter increased by one standard deviation,
e.g., 11.53 $\mu g/m^3$ to 18.50 $\mu g/m^3$, stock returns on that day were 11.9 percent
lower. Additionally, they found that increases in particulate matter led to
increases in the CBOE Volatility Index (VIX), a measure of fear and stock
market volatility. Both the decline in returns and the increase in the VIX
suggest that air pollution increased investors' risk aversion.

Lepori examined air pollution and stock returns through a natural experiment in Italy.[37] On April 15, 1994, the floor trading system was abandoned on
Italy's Milan Stock Exchange. Prior to this date, floor traders worked on the
exchange floor and were subject to the air pollution of the city. Afterwards,
technology replaced these floor traders so that they could operate remotely. As
such, Lepori examined the influence of air pollution on the stock returns before
and after 1994. Once the traders were no longer on the exchange floor, the air
pollution should have had a smaller effect on returns. The study confirmed this
relationship and reported that from 1989 to April 14, 1994, increases in air
pollution were associated with lower returns on the Milan Stock Exchange.
After April 14, 1994, the Milan air pollution had no impact on returns.

Huang and colleagues examined the influence of pollution on investor performance using portfolio data from over 80,000 Chinese traders from 34 cities.[38]
The authors matched the air pollution index in the 34 cities with the investors
that lived in the associated city. The study found that people made worse
investment decisions on days when air pollution was poor. Figure 10.2 illustrates

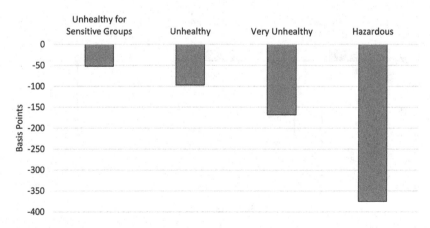

Figure 10.2 Underperformance of Stocks Based on Air Pollution on Date of Purchase

the difference between annualized returns on stocks purchased on healthy air days compared to unhealthy air days. For example, a stock purchased on a day when the air pollution index was unhealthy for sensitive groups had an annualized stock return 52 basis points lower than stocks purchased on days that had healthy levels of pollution. When air pollution was at its worst hazardous level, the stocks purchased underperformed stocks purchased in healthy levels of pollution by 375 basis points. This result suggests that investors make worse financial decisions and purchase poor stocks when air pollution is unhealthy.

Li and colleagues conducted a similar study examining the disposition effect.[39] The disposition effect is the behavior of investors selling their winning positions too soon and holding their losers too long. This is driven by the positive feelings of pride when a winner is sold and avoiding the negative feelings of regret when a loser is sold. This behavior is irrational for three reasons. First the winner sold continues to perform well, so it was sold too soon. Second, the loser being held continues to underperform, so it should have been sold. Finally, capital gains tax policy requires taxes to be paid for selling a winner at a capital gain, but a reduction in taxes is likely when a loser is sold for a capital loss. In their study, the authors obtained the trading history of nearly 775,000 brokerage accounts spanning more than 200 cities. The study matched the air pollution index to where each of the traders lived. Consistent with the idea that air pollution can both cause cognitive deficits and change risk preference, the authors found that investors were more likely to commit the disposition effect when air pollution was higher.

Allergies and Financial Decisions

Allergies are caused by an external substance such as peanut butter or pollen. Similarly to pollution, allergies can cause short-term health and cognitive deficits. Pantzalis and Ucar examined the effect of allergies on investor decision making.[40] The authors matched pollen counts to the city in which a firm was headquartered. Note that there is a bias in which investors tend to buy the stock of their local firms. This "local effect" version of the familiarity bias means that investors living closer to a firm's headquarters are more likely to buy and sell shares of that company. Thus, pollen near the company headquarters may influence many of the investors of that stock. The study found that as pollen increased, there were lower trading volumes, lower stock returns, and under-reactions to earnings announcements. This suggests that investors were distracted by the effect of allergies on their cognitive state.

Natural Disasters and Financial Decision Making

Natural Disasters and Behavior

Researchers have also investigated the impact of natural disasters on financial decision making. Natural disasters can have an adverse effect on health and community resources. Additionally, they can change risk preferences, possibly through emotions. For example, consider that someone had job offers in both Los Angeles and New York City. Just as they were about to make a decision, there was a large earthquake in Los Angeles. This natural disaster could influence the person's risk preference to prefer New York City instead of Los Angeles. Similarly, researchers argue that being exposed to more extreme events (such as large-scale hurricanes) may make people more sensitive to the consequences of risk and cause them to become more risk averse. However, it may also be possible that living through this event provides them with a better ability to handle risk, thereby allowing them to take on more risk.

Natural Disasters and Financial Decisions

Bernile and colleagues examined the impact of someone having lived through a disaster in his or her youth and their subsequent behavior as a chief executive officer (CEO).[41] The authors matched natural disasters to locations where the CEO lived when he or she was aged between five and 15 years old. They used the Sheuldus database which included earthquakes, volcanic eruptions, tsunamis, hurricanes, tornadoes, severe storms, floods, landslides, and fires. The study then sorted the natural disasters into three groups based on the number of people who died in the natural disaster: no fatality, medium fatality, and extreme fatality. They found that about 33 percent of CEOs experienced no natural disasters with fatalities, 56 percent experienced a medium fatality natural disaster, and 11 percent experienced an extreme fatality natural disaster. The authors then examined how the natural disaster events influenced risk taking by the CEO through the firms' financial leverage and cash-to-asset ratios. They found that the firms of CEOs who had experienced a medium fatality natural disaster had 3.3 percent higher leverage ratios and 1.2 percent less cash per dollar of assets than firms whose CEOs had experienced no natural disasters with fatalities. These CEOs are taking more risk. However, firms of CEOs who experienced an extreme fatality natural disaster displayed 3.5 percent lower financial leverage and 2.2 percent more cash per dollar of assets than firms of CEOs who did not experience a natural disaster with fatalities. These CEOs are taking less risk. Together, the study suggested that there is an inverse U-shaped relationship between fatalities in natural disasters experienced in adolescence and financial risk taking as a CEO.

Dessaint and Matray examined managers' reactions to salient risk following a hurricane strike.[42] One of the more common behavioral biases is to overweigh the probability of events that have a small risk of occurring, such as winning the lottery, or a large-scale hurricane hitting the same area again in the near future. The authors found that managers of firms that were unaffected by a hurricane, but who were located quite close to the affected area, responded to the possibility of a hurricane taking place in the near future by increasing cash holdings during the next year. These managers also mentioned hurricanes in their subsequent public accounting statements (10-Ks and 10-Qs) as posing a potential risk to the firm. In fact, hurricanes were mentioned as a risk 86 percent more times than the probability of their occurrence. However, after the increase in cash holdings over the first year, these firms lowered their cash holdings back to pre-hurricane levels within the following year (i.e., year 2). This suggests that as the length of time increases from the occurrence of the hurricane and the salient risk subsides, the managers forgot about the hurricane. Overall, the study showed that managers committed the behavioral bias of overweighting small probabilities, which resulted in higher cash holdings and fewer investments for the firm.

Bennett and Wang examined the impact of natural disasters in the municipal bond market using the Sheldus database.[43] The authors argued that following natural disasters, investors may be less likely to invest in the impacted region due to the risk of another natural disaster damaging their investment. If they were to invest in the area, they might require a higher rate of return to compensate for the perceived increased risk. The study found that following a large natural disaster, municipal bond yields in the affected counties increased in the first three months consistent with the salient risk conjecture. Additionally, there was a lower demand for those municipal bonds. However, the credit rating companies did not overestimate the probability of another large-scale natural disaster and did not change the countries' credit ratings. The issuers of municipal bonds also behaved differently following a natural disaster. They tended to issue bonds with shorter maturities than usual to help to finance the rebuilding resulting from the natural disaster.

Terrorism and the Stock Market

Terrorism and Behavior

There are very few events in our lives when we remember exactly where we were at an exact moment in time. Many of these are happy events, such as a wedding, the birth of a child, or the first day of a new job. However, there are also sad events that are forever ingrained into our brains, such as terrorist attacks.

Terrorist attacks are designed to create terror. Even if you don't know anyone directly affected, the attack can generate intense negative emotions such as anger, sadness, fear, anxiety, and despair. Fear and stress after a terrorist attack can adversely affect a person's cognitive ability to retrieve and analyze investment information, thus making complex trading tasks more difficult. Ahern illustrated that a terrorist attack in one country impacts the economic preferences of people from that country but living in another country.[44] These people experienced a substantial decrease in trust and subjective well-being.

Terrorism and the Stock Market

Wang and Young examined the impact of terrorist attacks on the risk aversion of investors.[45] This study used a database that reported all 409 attacks globally from 1984 to 2010. They matched this data with inflows and outflows of equity and bond mutual funds. The authors found that following terrorist attacks, investors shift their portfolios in order to reduce risk. They concluded that investors:

- sold equity mutual funds,
- purchased bond mutual funds, and
- shifted equity investments from small cap mutual funds to large cap funds.

Agarwal and colleagues used the 2008 Mumbai terrorist attacks as a natural experiment to study how the extreme shock affected people's trading activity and performance.[46] The attacks resulted in hundreds of fatalities and caused tremendous panic and fear among the general public. The attacks are referred to as "India's 9/11." The authors analyzed investor behavior after the attacks utilizing a dataset containing all trading records on the National Stock Exchange of India. The study found significantly less trading by Mumbai investors compared with investors located outside of Mumbai. Interestingly, the reduced trading activity did not begin right away but on the fourth day after the attacks. In addition to trading less, the Mumbai traders performed worse after the attacks. The authors concluded that traders' cognitive abilities were so impaired that they were led to make poor trading decisions.

Summary

Environmental factors such as the weather, air pollution, natural disasters, and terrorist attacks can have adverse effects on health, cognition, and mood that then influence financial decisions. Specifically, sunshine can lead to a positive mood that causes the stock market to rise and investors to take more risk and to be more optimistic. Conversely, a negative mood caused by SAD and the lack of daylight causes less risk taking, more pessimistic views, lower stock returns, lower volatility, and higher bid-ask spreads. Pollution also has a significant impact on the stock market and is associated with lower stock returns, higher volatility, and poor financial decision making. Allergens like pollen also cause cognition problems for investors. In summary, natural disasters and terrorist attacks cause investors to be more risk averse in their behavior.

Questions

1   How does the weather affect mood?
2   How does the weather impact financial decisions?
3   How does air pollution effect the brain?
4   How does air pollution affect the stock market?

Notes

1 E. Howarth and M. S. Hoffman. 1984. A multidimensional approach to the relationship between mood and weather. *British Journal of Psychology* 75:1, 15–23.

2 Aniko Hannak, Eric Anderson, Lisa Feldman Barrett, Sune Lehmann, Alan Mislove, and Mirek Riedewald. 2012. Tweetin' in the rain: Exploring societal-scale effects of weather on mood. In Sixth International AAAI Conference on Weblogs and Social Media, May.

3 Jeffrey L. Sanders and Mary S. Brizzolara. 1982. Relationships between weather and mood. *Journal of General Psychology* 107:1, 155–156.

4 Bruce Rind and David Strohmetz. 2001. Effect of beliefs about future weather conditions on restaurant tipping. *Journal of Applied Social Psychology* 31:10, 2160–2164.

5 John M. Eagles. 1994. The relationship between mood and daily hours of sunlight in rapid cycling bipolar illness. *Biological Psychiatry* 36:6, 422–424.

6 Glen H. Tietjen and Daniel F. Kripke. 1994. Suicides in California (1968–1977): Absence of seasonality in Los Angeles and Sacramento counties. *Psychiatry Research* 53:2, 161–172.

7 Anna Bassi, Ricccardo Colacito, and Paolo Fulghieri. 2013. 'O sole mio: An experimental analysis of weather and risk attitudes in financial decisions. *Review of Financial Studies* 26:7, 1824–1852.

8 David Hirshleifer and Tyler Shumway. 2003. Good day sunshine: Stock returns and the weather. *Journal of Finance* 58:3, 1009–1032.

9 William N. Goetzmann, Dasol Kim, Alok Kumar, and Qin Wang. 2014. Weather-induced mood, institutional investors, and stock returns. *Review of Financial Studies* 28:1, 73–111.

10 Carolina Johansson, Christina Smedh, Timo Partonen, Petra Pekkarinen, Tiina Paunio, Jenny Ekholm, Leena Peltonen, Dirk Lichtermann, Juni Palmgren, Rolf Adolfsson, and Martin Schalling. 2001. Seasonal affective disorder and serotonin-related polymorphisms. *Neurobiology of Disease* 8:2, 351–357.

11 Raymond W. Lam and Robert D. Levitan. 2000. Pathophysiology of seasonal affective disorder: A review. *Journal of Psychiatry and Neuroscience* 25:5, 469.

12 Norman E. Rosenthal, David A. Sack, J. Christian Gillin, Alfred J. Lewy, Frederick K. Goodwin, Yolande Davenport, Peter S. Mueller, David A. Newsome, and Thomas A. Wehr. 1984. Seasonal affective disorder: A description of the syndrome and preliminary findings with light therapy. *Archives of General Psychiatry* 41:1, 72–80.

13 Robert M. Cohen, Michael Gross, Thomas E. Nordahl, William E. Semple, Dan A. Oren, and Norman Rosenthal. 1992. Preliminary data on the metabolic brain pattern of patients with winter seasonal affective disorder. *Archives of General Psychiatry* 49:7, 545–552.

14 Jeanne Molin, Erling Mellerup, Tom Bolwig, Thomas Scheike, and Henrik Dam. 1996. The influence of climate on development of winter depression. *Journal of Affective Disorders* 37:2–3, 151–155.

15 Lisa A. Kramer and J. Mark Weber. 2012. This is your portfolio on winter: Seasonal affective disorder and risk aversion in financial decision making. *Social Psychological and Personality Science* 3:2, 193–199.

16 Guy Kaplanski, Haim Levy, Chris Veld, and Yulia Veld-Merkoulova. 2015. Do happy people make optimistic investors? *Journal of Financial and Quantitative Analysis* 50:1–2, 145–168.

17 Mark J. Kamstra, Lisa A. Kramer, and Maurice D. Levi. 2003. Winter blues: A SAD stock market cycle. *American Economic Review* 93:1, 324–343.

18 Ian Garrett, Mark J. Kamstra, and Lisa A. Kramer. 2005. Winter blues and time variation in the price of risk. *Journal of Empirical Finance* 12:2, 291–316.

19 NYSE Internal Database and Consolidated Tape Statistics. 2019. Retrieved on March 31, 2019 from www.nyse.com/markets/nyse#marketquality.
20 Robert A. Clark, John J. McConnell, and Manoj Singh. 1992. Seasonalities in NYSE bid-ask spreads and stock returns in January. *The Journal of Finance* 47:5, 1999–2014; Richard D. Fortin, R. Corwin Grube, and O. Maurice Joy. 1989. Seasonality in NASDAQ dealer spreads. *Journal of Financial and Quantitative Analysis* 24:3, 395–407.
21 Hans R. Stoll. 1978. The supply of dealer services in securities markets. *Journal of Finance* 33:4, 1133–1151.
22 Ramon P. DeGennaro, Mark J. Kamstra, and Lisa A. Kramer. 2008. Does risk aversion vary during the year? Evidence from bid-ask spreads. Working Paper, University of York.
23 Mark J. Kamstra, Lisa A. Kramer, Maurice D. Levi, and Russ Wermers. 2017. Seasonal asset allocation: Evidence from mutual fund flows. *Journal of Financial and Quantitative Analysis* 52:1, 71–109.
24 Lazaros Symeonidis, George Daskalakis, and Raphael N. Markellos. 2010. Does the weather affect stock market volatility? *Finance Research Letters* 7:4, 214–223.
25 Steven D. Dolvin, Mark K. Pyles, and Qun Wu. 2009. Analysts get SAD too: The effect of seasonal affective disorder on stock analysts' earnings estimates. *Journal of Behavioral Finance* 10:4, 214–225.
26 Kin Lo and Serena Shuo Wu. 2017. The impact of seasonal affective disorder on financial analysts. *Accounting Review* 93:4, 309–333.
27 Steven D. Dolvin and Mark K. Pyles. 2007. Seasonal affective disorder and the pricing of IPOs. *Review of Accounting and Finance* 6:2, 214–228.
28 Lucio G. Costa, Toby B. Cole, Jacki Coburn, Yu-Chi Chang, Khoi Dao, and Pamela Roque. 2014. Neurotoxicants are in the air: Convergence of human, animal, and in vitro studies on the effects of air pollution on the brain. *BioMed Research International* 2014:736385, 1–8.
29 Lotte Risom, Peter Møller, and Steffen Loft. 2005. Oxidative stress-induced DNA damage by particulate air pollution. Mutation *Research/Fundamental and Molecular Mechanisms of Mutagenesis* 592:1–2, 119–137.
30 Kenneth Y. Chay and Michael Greenstone. 2003. The impact of air pollution on infant mortality: Evidence from geographic variation in pollution shocks induced by a recession. *The Quarterly Journal of Economics* 118:3, 1121–1167.
31 Aaron J. Cohen, H. Ross Anderson, Bart Ostra, Kiran Dev Pandey, Michal Krzyzanowski, Nino Künzli, Kersten Gutschmidt, Arden Pope, Isabelle Romieu, Jonathan M. Samet, and Kirk Smith. 2005. The global burden of disease due to outdoor air pollution. *Journal of Toxicology and Environmental Health* 68:13–14, 1301–1307.
32 Xin Zhang, Xi Chen, and Xiaobo Zhang. 2018. The impact of exposure to air pollution on cognitive performance. *Proceedings of the National Academy of Sciences* 115:37, 9193–9197.
33 Yulong Chen. 2018. Air pollution and academic performance: Evidence from China. Working Paper, Iowa State University.
34 Sefi Roth. 2016. The contemporaneous effect of indoor air pollution on cognitive performance: Evidence from the UK. Working Paper, London School of Economics.
35 Tamir Levy and Joseph Yagil. 2011. Air pollution and stock returns in the US. *Journal of Economic Psychology* 32:3, 374–383.
36 Anthony Heyes, Matthew Neidell, and Soodeh Saberian. 2016. The effect of air pollution on investor behavior: Evidence from the S&P 500. NBER Working Paper No. 22753.
37 Gabriele M. Lepori. 2016. Air pollution and stock returns: Evidence from a natural experiment. *Journal of Empirical Finance* 35:January, 25–42.

38 Jiekun Huang, Nianhang Xu, and Honghai Yu. 2017. Pollution and performance: Do investors make worse trades on hazy days? Available at SSRN 2846165.

39 Jennifer J. Li, Massimo Massa, Hong Zhang, and Jian Zhang. 2017. Behavioral bias in haze: evidence from air pollution and the disposition effect in China. SSRN Working Paper 2993763.

40 Christos Pantzalis and Erdem Ucar. 2018. Allergy onset and local investor distraction. *Journal of Banking & Finance* 92:July, 115–129.

41 Gennaro Bernile, Vineet Bhagwat, and P. Raghavendra Rau. 2017. What doesn't kill you will only make you more risk-loving: Early-life disasters and CEO behavior. *Journal of Finance* 72:1, 167–206.

42 Olivier Dessaint and Adrien Matray. 2017. Do managers overreact to salient risks? Evidence from hurricane strikes. *Journal of Financial Economics* 126:1, 97–121.

43 Benjamin Bennett and Zexi Wang. 2019. Costs of natural disasters in public financing. Fisher College of Business Working Paper No. 2019-03-009.

44 Kenneth R. Ahern. 2018. The importance of psychology in economics activity: Evidence from terrorist attacks. NBER Working Paper No. 24331.

45 Albert Y. Wang and Michael Young. Forthcoming. Terrorist attacks and investor risk preference: Evidence from mutual fund flows. *Journal of Financial Economics*.

46 Vikas Agarwal, Pulak Ghosh, and Haibei Zhao. 2019. Extreme stress and investor behavior: Evidence from a natural experiment. Working paper, Lehigh University, Pennsylvania, December.

11 The Personality of a Successful Investor

Personality is "a set of distinctive traits and characteristics."[1] Some people are familiar with the "big five" personality traits that include extroversion, agreeableness, neuroticism, openness, and conscientiousness. Each personality is made up of various amounts of each trait. Each trait is measured as a continuum between two extremes. For example, each person's personality has some of the extroversion trait that can be characterized as being somewhere between extremely extroverted to extremely introverted. Scholars have also determined that certain combinations of personality traits are essential to specific careers fields, like finance. Specifically, personalities with strong traits in narcissism, Machiavellianism, and psychopathy, known as the "dark triad," are common in the financial industry. There are also noncognitive abilities/traits that impact decision making. For example, traits such as self-efficacy and self-control are characteristics that are not usually encompassed by measures of cognitive ability or personality. Yet they can be critical in the making and execution of decisions. This chapter illustrates how these personality traits and noncognitive abilities relate to financial decisions.

Big Five Personality Traits and Financial Decisions

Big Five Personality Traits

Table 11.1 presents the big five personality traits and several characteristics that describe each trait, according to Costa and McCrae.[2] A person high in agreeableness is someone who trusts easily, is straightforward, has concern for others, would rather forgive and forget, is humble, and is sympathetic. If these characteristics do not describe a person, then that person scores low in agreeableness. Conscientiousness refers to people who are well-prepared, organized, ethical, reliable, are diligent workers, have self-discipline, and think carefully before acting. Extroversion refers to people having traits of being sociable, outgoing, friendly, cheerful, and optimistic. People with these traits are often group leaders. Neuroticism refers to people who are anxious, easily frustrated, sad, self-conscious, impulsive, and vulnerable to stress. Finally, openness refers to people who have a vivid imagination, appreciate

Table 11.1 Characteristics of the Big Five Personality Traits

Agreeableness	Conscientiousness	Extroversion	Neuroticism	Openness
Altruism	Achievement Striving	Assertiveness	Angry Hostility	Actions
Compliance	Deliberation	Excitement-Seeking	Anxiety	Aesthetics
Modes	Dutifulness	High Energy	Depression	Feelings
Straightfor-wardness	Organized	Positive Emotions	Impulsiveness	Ideas
Sympathetic	Self-Discipline	Preference for large groups	Self-Consciousness	Values
Trust	Well Prepared	Warmth	Vulnerability	Vivid Imagination

art, poems, and beauty, are receptive to their own feelings, are willing to try new things such as activities, places, or unusual foods, are intellectually curious, and question the world. As you think about these personality traits, it may come to mind that some people may be well suited to making financial decisions. Someone who scores high on conscientiousness (highly prepared and organized, self-disciplined, and strives for achievement) may make a better financial trader than someone who has a high level of neuroticism (anxious, sad, and impulsive).

Big Five Personality Traits and Neural Activity

Have you ever tried to change something about your personality? Trying to be more outgoing, organized, or less anxious is really difficult. One reason for this is that your personality may be hard-wired into your brain. So, making a change is to literally rewire your brain. Using functional magnetic resonance imaging (fMRI), researchers have found that neuroticism, extroversion, openness, conscientiousness, and agreeableness traits show significant relationships to activity in the prefrontal cortex, cingulate cortex, temporal lobe, occipital lobe, striatum, and amygdala during a variety of tasks (See Chapter 5 for a discussion on these brain regions).[3] Researchers also find that fundamental differences in personality are related to the neuron volume in different regions of the brain, including the prefrontal cortex, hippocampus, and cerebellum. Having increased volume in these brain regions may improve function in these areas because increases in neurons can create more significant electrochemical outputs. As such, it is possible that personality traits can be related to improved cognitive function, memory, decision making, and emotion.[4]

Other researchers explore the relationship between personality traits and neurotransmitters and hormones. For example, extroversion is related to higher levels of dopamine, a neurotransmitter that modulates cognitive and emotional processes and is linked to increased risk taking.[5] Neuroticism is related to higher levels of the neurotransmitter norepinephrine[6] and cortisol,[7] which both suggest lower risk taking (See Chapter 6 on hormones).

Big Five Personality Traits and Experimental Risk Taking

Nicholson and colleagues examined personality and risk taking in relation to many social factors, specifically using the categories recreation, health, career, finance, safety, and social.[8] The study found that people who scored highly on extroversion and openness and lower on neuroticism, agreeableness, and conscientiousness exhibited more risk taking associated with these factors. Prinz and colleagues examined the influence of personality traits and risk taking in a Holt and Laury lottery task.[9] During this task participants were given ten possible decisions where option A was a chance to win $2 or $1.60 with varying probabilities and option B was a chance to win $3.85 or $0.10 with varying probabilities. In addition to environmental factors such as family wealth, the authors found that people who had higher scores on openness and lower scores on agreeableness were more willing to take risk.

Filbeck and colleagues examined the relationship between financial risk aversion and personality within the context of expected utility theory.[10] Expected utility theory measures the risk preferences of people when facing uncertainty. Specifically, the value associated with a particular gamble is assumed to be the statistical expectation of the gamble, which is the sum of the probability of each outcome multiplied by the value of each outcome. Note that the expected value of a choice between a 50 percent chance to win $100 and a 50 percent chance to win $0 is $50 (= 0.5 × $100 + 0.5 × $0). However, an even chance choice between payoffs of $200 and −$100 is also $50. This second choice has much more variance or risk. Also, a choice with a 10 percent chance of winning $500 and a 90 percent chance of winning $0 has an expected value of $50 with a more skewed set of outcomes. Through a survey that asked the subjects various questions about the amount of risk they were willing to take in order to meet their future retirement needs they could assess their desired risk level as measured by variance and skewness preferences. The authors found that people who scored lower on agreeableness and openness and higher on conscientiousness were more willing to take risk.

Oehler and Wedlich conducted an experiment to explore the relationship between personality traits and risk attitude, risk perception, and return expectations.[11] To measure risk attitude, they conducted a study that provided subjects with either a guaranteed amount between $1 and $9 or a gamble that had various probabilities with an outcome of $10. The study assessed risk perception and return expectations by showing the subjects a five-year chart of stock price movements for three stocks and asked them to predict the next year's price and asked how risky they believed the stock would be in the next year. The results showed that higher scores on extroversion and lower scores on neuroticism were related to taking more risk in a financial setting. Furthermore, higher scores on conscientiousness were related to being more risk averse and to viewing investments as being more risky.

Durand and colleagues examined the impact of personality and gender on myopic loss aversion.[12] Myopic loss aversion depends on two behavioral biases: loss aversion and mental accounting. Loss aversion refers to the fact that people are more sensitive to losses than they are to gains (i.e., losing $50 feels worse in magnitude, while winning $50 feels good in magnitude). Indeed, people behave in seemly irrational ways in order to avoid taking a loss. Mental accounting is the cognitive bias on which people tend to consider each investment individually rather than all together as a portfolio. People examine the gain/loss position of each stock, mutual fund, and retirement account separately. Putting the two ideas together, *myopic loss aversion* is simply the motivation from these two tendencies to avoid losses in each individual investment account. In this study, the authors endowed subjects with $100, each of whom then had to decide whether to gamble between $0 and $100 in many different rounds. The gamble involved three different colored balls (red, blue, and green). If the subject's color was pulled from the bag, they earned 2.5 times their gamble; if their colored ball was not pulled out, they lost their gamble. The subjects divided into two groups denoted by a frequent or an infrequent treatment. In the frequent treatment, the subject played nine rounds of this gamble one by one. In the infrequent treatment, the subject still played nine rounds; however, the only decisions they made were before rounds one, four, and seven. The amount they decided to bet in round one would be the same bet in round two and three, and so on, for the further rounds. The idea behind myopic loss aversion is that the less frequently investors check and rebalance their portfolio, the lower level of emotion they will experience when trading and thus be more willing to take risk (i.e., gamble more). Therefore, the infrequent group should take more risk. Additionally, there is a plethora of research that shows that women tend to be more risk averse in social and financial situations, which suggests that women are more susceptible to myopic loss aversion than men. The initial finding was that women, in fact, were more likely to commit the myopic loss aversion bias as predicted. However, after considering the big five personality traits, the authors found that the real cause of myopic loss aversion had little to do with gender and more to do with personality traits. In other words, lower extroversion scores and higher neuroticism scores were associated with myopic loss aversion. As such, gender differences in risk taking (See Chapter 4 on gender) may have less to do with differences in biology or societal norms but more to do with fundamental traits such as personality.

Big Five Personality Traits and Experimental Investments

Durand and colleagues examined the impact of personality traits on investment decisions.[13] A common instructional tool for classes in investment at universities is to have students participate in a stock market game. In this game, the students spend the entire semester managing a portfolio of all stocks in the stock market with pretend money. The authors used this game to reveal student performance and investment decisions and then matched those decisions with student personality data. In their first study, the authors found the following results:

- Higher scores on extroversion and lower scores on agreeableness were related to greater exposure to equities in portfolios.
- Higher scores on extroversion were related to higher portfolio returns.
- Higher scores on extroversion and lower scores on conscientiousness were related to higher Sharpe ratios (risk adjusted return).
- Lower levels of extroversion were related to less trading and lower portfolio turnover.

In a second experiment, Durand and colleagues found the following results:[14]

- Lower scores on openness were related to higher portfolio diversification.
- Lower scores on agreeableness and conscientiousness were related to buying stocks with higher recent returns.
- Higher scores on agreeableness and conscientiousness were positively related to the disposition effect.

In their third paper, Durand and colleagues examined a multitude of measures of overconfidence. Overall, they found that higher scores on openness and agreeableness and lower scores on extroversion were related to being more overconfident and exhibiting more overreaction to news.[15]

Oehler and colleagues conducted an experiment in which subjects participated in a 15-period trading game.[16] At the beginning of the experiment, participants were endowed with $3,000 and had to invest in five identical stocks that had an equal probability of earning $0, $8, $28, or $60 per period. The subjects were required to buy and sell over the same period as all of the other participants. Additionally, each stock received the same dividend in each period and had an expected dividend of $24 (= (0 + 8 + 28 + 60) ÷ 4). Unused cash earned 1.5 percent interest, which resulted in an expected price of $1,600 per stock. The authors found that people who scored higher on extroversion were more willing to buy the stock at a premium (over $1,600), while individuals who scored higher on neuroticism were less likely to hold the risky asset.

Big Five Personality Traits and Investments

Tauni and colleagues examined the influence of personality traits as a mediator of financial advice and trading behavior.[17] The authors sent an online questionnaire to the trading accounts of investors in China. The authors asked questions to determine personality traits, trading behavior, and where investors found their financial advice. The authors found that investors who gained financial advice from more reliable sources were more likely to make buy and sell transactions, while investors were less likely to trade when they obtained advice via word of mouth. Investors were more likely to trust reliable sources. However, investors who scored higher on extroversion and agreeableness were more likely to trade on any information regardless of the source compared to investors who scored higher on the conscientiousness trait.

Bucciol and Zarri used the U.S. Health and Retirement Study (HRS) to examine the link between personality traits and investor portfolios.[18] The HRS followed about 7,000 households with heads of households aged between 51 and 61 years and asked them many questions, including about their personalities and retirement portfolios. The authors found that people with higher scores for agreeableness and neuroticism held portfolios with less risk, including holding lower percentages of equity.

Big Five Personality Traits and Corporate Finance

Gow and colleagues studied the influence of personality traits on company policies.[19] The study found that a chief executive officer (CEO)'s personality was related to the company's behavior. Specifically, higher CEO scores on openness were related to higher company investment in research and development and lower financial leverage. The authors argued that openness was related to being creative, which could influence the CEO to try to use that creativity to innovate and develop new products. They also found that higher scores on conscientiousness were related to higher book-to-market ratios suggesting that a CEO's preference for rules and low adaptability resulted in low growth for the firm (i.e., higher book-to-market ratios). Furthermore, extroversion was associated with lower returns on assets suggesting that extroversion may not be a positive trait in business due to its link to overconfidence, while openness was inversely related to profitability.

Hrazdil and colleagues used machine learning and artificial intelligence from IBM on transcripts from the question and answer portion of company conference calls to measure the personality traits of the firm's CEO.[20] The study found that openness and extroversion were positively related to higher audit fees. Higher audit fees are typically due to firms having higher amounts of risk suggesting that the extroversion and openness of the CEO were related to the firm having higher risk.

Liu examined the influence of personality traits of the company's CEO and stock price crash risk.[21] The authors defined stock price crash risk as when managers hold onto bad news and fail to report it to the public. This hoarding of bad news temporarily keeps the stock price from declining while the bad news piles up. When the bad news leaks out all at once the stock price crashes. Previous research suggests that items such as earnings management, earnings smoothness, and tax avoidance also play a role in stock price crash risk. The study found that CEOs who scored higher on the traits of agreeableness and neuroticism and lower on conscientiousness were more likely to withhold bad news, thus creating an increase in stock price crash risk.

The personality of the CEO can also have an effect outside of the firm. Becker and colleagues examined the influence of the CEO's personality on financial analysts' forecasts.[22] Since research suggests that extroverted people are more sociable and optimistic, the authors hypothesized that financial analysts would have biased earnings forecasts because an extroverted CEO's optimism clouded the analysts' judgment. Consistent with this insight, the authors found that financial analysts issued higher earnings forecasts for companies with an extroverted CEO and that this biased forecast got larger as the forecast horizon increased.

Big Five Personality Traits and Personal Finances

Brown and Taylor used data from the British Household Panel Survey to examine the influence of personality traits on household finances.[23] The authors looked at personal finances, such as how much debt the participants held and their use of personal loans or credit cards. The study found that people with higher levels of extroversion and openness and lower levels of conscientiousness tended to have higher amounts of debt.

Parise and Peijnenburg examined the influence of personality traits on financial distress.[24] The authors used data from the Longitudinal Internet Study for the Social Sciences from Tilburg University in the Netherlands. This dataset surveyed people annually about a multitude of subjects. Data on personality traits were matched to financial distress variables such as delinquencies in rent/mortgage, car loan, utilities, or other bills. The authors found that people who had higher scores on conscientiousness and lower scores on neuroticism were less likely to have delinquent debt. The authors argued that people with higher conscientiousness and lower neuroticism scores were more emotionally stable and well prepared. This allowed for better financial decisions and avoidance of disastrous financial situations.

Zamarro examined the impact of the big five personality traits and the noncognitive trait of grittiness on retirement preparation and financial capability.[25] The author defined grittiness as "perseverance and passion for long-term goals." The study provided respondents with a 20-question survey on financial literacy and also asked about their total net worth in order to measure financial capability. The study used a survey from the Consumer Financial Protection Bureau that asked respondents how well they could meet current and future financial obligations to measure financial well-being. The third variable of interest regarded their preparation for retirement, whereby the respondents were asked questions about how they were preparing for their retirement. The study found the following:

- Openness was positively related to financial literacy.
- Conscientiousness and extroversion were positively related to perceived financial literacy, while agreeableness and neuroticism were inversely related to financial literacy.
- Conscientiousness, extroversion, neuroticism, and grittiness were positively related to better financial well-being and credit scores.
- Conscientiousness, extroversion, and grittiness were positively related to retirement preparation.

The Dark Triad and Financial Decisions

The Dark Triad

The dark triad refers to the personality traits of narcissism, Machiavellianism, and psychopathy. While all three have similarities, narcissism is characterized by pride and egotism. Machiavellianism is characterized as being manipulative and focused on self-interest. Psychopathy is characterized as being antisocial, narcissistic, Machiavellian, impulsive, callous, and remorseless. The three traits are called the dark triad because they are malevolent qualities. Additionally, these traits seem to be important to business and the financial industry.

The Dark Triad and Business

Dark triad traits are more common in business than in most other occupations. For example, true psychopaths account for about 1 percent of the population in the United States. However, approximately 4 percent of business executives are psychopaths.[26] Additionally, Christopher Bayer, who is a psychologist who provides care to professionals on Wall Street, believes that the percentage of psychopaths working on Wall Street is closer to 10 percent.[27] Consistent with this idea, Robert Hare, a leading researcher in psychopathy, says that "If [he] wasn't studying psychopaths in prison, [he'd] do it at the stock exchange."[28] Furthermore, researchers found that people who commit white-collar crimes such as Ponzi schemes, embezzlement, or insider trading, score higher on psychopathy tests than do violent criminals.[29]

Researchers found that business majors have more psychopathic tendencies than health profession majors[30] and score higher on the traits of narcissism, Machiavellianism, and psychopathy than do psychology, law, and political science students.[31] Additionally, studies found that people who pursue business careers score higher on narcissism and psychopathy traits than those having other careers.[32] Why are people who score higher on the dark triad more likely to pursue business careers? Some researchers argue that the business curriculum may cause unethical behavior.[33] For example, Schneider and Prasso argued that the business curriculum caused students to shift away from the view that companies should create products and services that were most beneficial for its customers toward a policy focusing on maximizing shareholder wealth.[34] Others argue that the prevailing culture in the financial industry may cause psychopathic behavior.[35]

Psychologists have distinguished between the successful and the unsuccessful psychopath. The definition of a successful psychopath is someone who displays psychopathic traits but who has been able to avoid prison and lead a successful lifestyle. In contrast to incarcerated psychopaths, these successful psychopaths have found a way to use their underlying psychopathic behaviors in a beneficial way.

So, why are the dark triad traits more prevalent in the business industry than in other occupations? Hill linked many psychopathic traits to the desirable skills of leaders and executives.[36] The author examined psychopathic tendencies that were synonymous with someone who could climb the corporate ladder to become an executive. For example, psychopathic tendencies that may transfer well to the corporate world include being strategically minded, being a visionary, having courage, being able to make tough decisions and live with them, being analytical, being comfortable taking risk, and being confident. Furthermore, psychopathic CEOs tend to be creative, strategic thinkers with excellent communication skills. However, they also tend to have poor cooperation and management abilities.[37] As such, if a psychopath can suppress some of the undesirable traits such as impulsiveness and poor cooperation, they may be able to use their personality to their best advantage.

The Underlying Conditions of the Dark Triad

The dark triad traits have been linked to neurological functions. For example, researchers found structural and functional differences in the amygdala and prefrontal cortex in psychopaths compared to normal people.[38] Additionally, research suggests that psychopathy is linked to a neurotransmitter imbalance.[39] These biological differences can cause impairments in the brain regions and cause people to become more risk seeking, show little fear, and have callous and socially deviant behavior.

In addition to the research that shows a biological explanation for psychopathy, researchers have also examined whether nurture can cause psychopathic behavior. Blanchard and Lyons found that negative parenting styles were related to psychopathic traits. There also appears to be gender differences. Having a controlling mother who avoided becoming emotionally attached (called avoidant attachment) was linked to psychopathy in men. There was little relationship to parenting styles in female psychopathy.

The Dark Triad and Financial Decisions

There is limited research on the impact of psychopathic behavior and financial decisions. Shank and colleagues examined the impact of psychopathic traits on financial risk and time preferences.[40] The authors used the Psychopathic Personality Inventory-Revised (PPI-R) survey on subjects and matched their scores with the Dynamic Experiments for Estimating Preferences to obtain their risk and time preferences. The risk decisions asked questions such as "would you prefer gamble A, which is a 50 percent chance to win $100 and a 50 percent chance of losing $15, or gamble B, which is a 90 percent chance to win $40 and a 10 percent chance of winning $10?" In this example, the expected value outcome is much higher for gamble A, which is $42.50 (= $0.5 \times \$100 + 0.5 \times -\$15$). The expected value of gamble B is only $37 (= $0.9 \times \$40 + 0.1 \times \$10$). However, gamble A involves the risk of losing money. This risk is not present in gamble B. Thus, this question asks whether you would be willing to give up $5.50 in this situation to guarantee that you do not lose money. A series of these types of questions were asked to assess the person's risk and loss aversion. A series of time decision questions asked, for example, "would you prefer to receive $250 today or $300 in one month?" The authors made two key findings:

- Psychopathy and some of its corresponding traits were related to making more rational financial decisions when it came to risk.
- Psychopathy and some of its traits were related to impulsive and irrational decisions when it came to time preferences and delaying payments.

Shank examined the influence of psychopathy on deceptive behavior in a large group of college students.[41] The students were given the PPI-R survey and then participated in two experiments to elicit deception. Additionally, the study used business students (who are said to be are more psychopathic than non-business students) and a group of health science majors. The first experiment was a cheap talk experiment in which the student was told the potential payout for option A and option B and was obliged to tell another student (i.e., their partner) which option was worth more in order for the second student to make a decision. For example, the first student was told that option A would earn them $5 and their partner $15, while option B would earn them $15 and their partner $5. Their partner would never be given any information other than that which the first student told them and would never know how much the actual payouts amounted to. As such, if the first student lied and told their partner that option B would earn the partner more money, the partner would never know that the student had lied. The author found that business students were more likely to lie to their partners in order to earn more money for themselves. Additionally, there was a positive relationship between psychopathy and sending a deceitful message to the partner. The second experiment involved an ethical views survey. This survey asked students the following question:

Mr. Johnson is about to close a deal and sell his car for $1,200. The engine's oil pump does not work well, and Mr. Johnson knows that if the buyer learns about this, he will have to reduce the price by $250 (the cost of fixing the pump). If Mr. Johnson does not tell the buyer, the engine will overheat on the first hot day, resulting in damages of $250 for the buyer. Being winter, the only way the buyer can learn about this now is if Mr. Johnson were to tell him. Otherwise, the buyer will learn about it only on the next hot day. Mr. Johnson chose not to tell the buyer about the problems with the oil pump.

In your opinion, Mr. Johnsons' behavior is completely fair, fair, unfair, very unfair.

After the students completed this scenario, they were asked to judge the following scenario with the same outcomes of completely fair, fair, unfair, very unfair:

What would your answer be if the cost of fixing the damage for the buyer in case Mr. Johnson does not tell him is $1,000 instead of $250?

In this question, the author found that business students were more likely to find that this deception was completely fair or fair compared to non-business students. Furthermore, psychopathic traits were also positively linked to believing that this situation was fair.

Given the link to the dark triad and unethical behavior, Ham and colleagues examined the influence of CFO narcissism and the quality of financial reporting.[42] The authors found that companies with narcissistic CFOs were more likely to use earnings management techniques, recognize losses in a less timely manner, have lower quality internal controls, and have a higher probability of financial statement restatements. These results show a consistent association between narcissistic personality and unethical executive behavior. Similarly, Aktas and colleagues examined the effect of CEO narcissism and the merger and acquisition process.[43] To measure narcissism, the authors used the prevalence of the CEO using personal pronouns such as "I" during his speeches. The authors found the following:

- Narcissistic CEOs earned higher bid premiums.
- Investors reacted worse to the takeover announcement of a targeted firm when the CEO had higher levels of narcissism.
- When acquiring firm CEOs measured higher on the narcissism scales, acquisition deals were initiated and negotiated quicker.
- Narcissism was linked to a lower probability of deal completion between both the target and acquirer CEO.

Brinke and colleagues examined the trading performance of hedge fund managers who displayed psychopathic behavior.[44] The authors watched interviews with hedge fund managers to determine verbal and nonverbal behaviors that were linked to the dark triad. They matched these personality traits to the funds' ten-year performance. The authors found that managers who are more psychopathic and narcissistic earned about 88 basis points less per year over the ten-year period. Additionally, the authors found that psychopathy and narcissism were linked to lower risk-adjusted returns (i.e., Sharpe ratio).

Noncognitive Abilities and Financial Decisions

Noncognitive Abilities

Noncognitive abilities can define a person just as much as personality traits. For example, if your company was hiring a new co-worker, would you rather have someone who is hard working and dedicated or someone who is lazy? As you would expect, many noncognitive abilities have been linked to being successful regardless of intelligence level. Researchers have found that noncognitive abilities are as important to social and economic success as are cognitive abilities.[45]

Heckman and Rubinstein examined noncognitive abilities, high school education, and job performance.[46] They argued that high school graduates, on average, had higher cognitive abilities than high school dropouts who earn their General Educational Development (GED). Additionally, those who earn their GED had, on average, higher cognitive abilities than high school dropouts who do not earn a GED. However, the literature shows that people who earn their GED earn lower wages than those who did not earn their GED. The authors found that high school dropouts who do not earn their GED had greater noncognitive abilities such as more work experience and less illicit drug use.

Noncognitive Abilities and Business

Brau and colleagues examined noncognitive abilities that are associated with being a finance major.[47] At a selective university, students could not declare a major until they enrolled. After their sophomore year, they could apply to be accepted into a major, which eliminated a selection bias as no student in the sample had been accepted into the business school. They examined differences in students who applied to be finance majors versus those who applied for any other major. The authors found that those applying to be a finance major had higher motivation, loftier career goals, and displayed a stronger work ethic than students who applied for non-finance majors.

Overconfidence, Optimism, and Financial Decisions

One of the most researched noncognitive traits is overconfidence—a trait that seems to be prevalent among corporate executives. Malmendier and Tate examined the relationship between overconfident CEOs and corporate investment.[48] They defined overconfidence by the timing of CEOs' decisions to exercise their executive compensation options and by their behavior of purchasing company stock. The authors argued that overconfident CEOs would exercise their options late and hold more company stock because the CEO believed that he or she could do great things with the company. If the company were to perform poorly, and the stock price to decline, it would negatively affect the CEO's personal portfolio and lower the value of their options. The authors found that overconfident CEOs overestimated the profitability of new projects that didn't live up to expectations. Due to overconfident CEOs overestimating their ability to earn high returns, Malmendier and Tate also examined firm acquisitions using the same measure of CEO overconfidence.[49] The authors reported that overconfident CEOs overpay for target companies, thereby destroying the benefits and value of the acquisition. Overconfident CEOs were also 65 percent more likely to acquire other firms. Finally, these acquisitions were poorly received by investors as the stock price for acquisitions from overconfident CEOs was 100 basis points lower than for firms with non-overconfident CEOs.

Kim and colleagues examined the impact of overconfident CEOs on stock price crash risk.[50] The authors argued that as overconfident CEOs continue to overestimate the profits of projects, this would result in poor long-term performance that would eventually lead to a stock price crash. Additionally, as overconfident CEOs held more stock than would most CEOs, they were less likely to discuss the poor performance of company projects as it would adversely affect their own portfolio. Consistent with these ideas, the authors found that companies with overconfident CEOs were more likely to experience a stock price crash than companies with CEOs who were not overconfident.

If having an overconfident CEO adversely affects the firm, why are they hired? Hirshleifer and colleagues asked this question in the context of firm innovations.[51] The authors found that firms with overconfident CEOs invested more in innovative projects, achieved more patents, and had more success with research and development divisions. Thus, it is possible that the overconfident CEO may be better for companies in certain industries.

Optimism and overconfidence are similar. Overconfidence tends to bring about a focus on one's own abilities, while optimism tends to focus on outcomes. Nevertheless, both may lead to similar business decisions. Heaton examined the impact of optimistic managers and corporate decisions.[52] When managers are overly optimistic, they overestimate the probability of projects having positive outcomes. Optimistic managers believe that outside investors undervalue their firm, i.e., the price is too low. This belief leads the manager to finance more projects internally, which tends to be more costly. That is, optimistic managers are more likely to decline external financing for good projects as they view the external funds as too costly, when it is actually the internal funding that is costly. Furthermore, similarly to overconfident managers, optimistic managers tend to overestimate the profitability of new projects.

Noncognitive Abilities and Financial Decisions

Kuhnen and Melzer examined the influence of the noncognitive trait of self-efficacy on financial delinquency using the National Longitudinal Survey of Youth (NLSY).[53] The NLSY provides surveys from participants' early childhood through adulthood and tracks many items including cognitive and noncognitive traits and financial experiences such as debt delinquency, savings, insurance purchases, credit applications, and retirement preparations. The authors found that more self-efficient people were less likely to have missed debt payments, to have assets repossessed, to have filed for bankruptcy, to have assets in collection, and to have been rejected for credit or loans. In addition, they were also more likely to have emergency funds in their savings accounts, to have purchased health insurance, and to be prepared for retirement.

Approximately 15 percent of college students who are offered interest-free loans through financial aid turn them down. Cadena and Keys examined why these college students might turn down zero interest loans using the National Postsecondary Student Aid Study (NPSAS).[54] The authors argued that students who had more self-control turned down loans in order to avoid the temptation of spending more money than was needed.

Some people believe that the stock market is just legalized gambling. Jadlow and Mowen examined similarities in noncognitive traits of investors and gamblers.[55] The authors provided an online survey that asked participants how well certain traits characterized them and provided questions about their investing and gambling habits. The authors found that gamblers and investors shared the traits of competitiveness, superstition, financial conservatism, quantitative ability, and a desire to make money. Additionally, gamblers were more likely to be able to delay gratification, be less emotionally stable, and be more impulsive, while investors displayed the opposite of these traits.

Summary

Personality traits and other noncognitive abilities play an important role in financial decisions. The personality trait of extroversion is consistently positively related to risk taking while neuroticism is inversely related to risk taking. Furthermore, extroversion is related to higher portfolio returns. Psychopathy and the other dark triad personality traits are prevalent in the financial industry and typically have negative consequences to firm performance. Finally, overconfidence and optimism can cause CEOs to make irrational decisions. Overall, a number of personality and other noncognitive traits are linked to successful financial decision making.

Questions

1 What are noncognitive traits? Give a few examples.
2 How do the big five personality traits relate to financial decisions of investors and corporate executives?
3 How do the dark triad of personality traits relate to the business industry and financial decisions?
4 How are overconfidence and optimism related to financial decisions?

Notes

1 "Personality." 2019. See Merriam-Webster.com. Retrieved July 1, 2019, from www.merriam-webster.com/dictionary/personality.
2 Paul T. Costa and Robert R. McCrae. 2009. The five-factor model and the NEO inventories. In James N. Butcher (Ed.), *Oxford Handbook of Personality Assessment*, New York, Oxford University Press, 299–322.
3 Mitzy Kennis, Arthur R. Rademaker, and Elbert Geuze. 2013. Neural correlates of personality: An integrative review. *Neuroscience & Biobehavioral Reviews* 37:1, 73–95.
4 Colin G. DeYoung, Jacob B. Hirsh, Matthew S. Shane, Xenophon Papademetris, Nallakkandi Rajeevan, and Jeremy R. Gray. 2010. Testing predictions from personality neuroscience brain structure and the big five. *Psychological Science* 21:6, 820–828.
5 Richard A. Depue and Paul F. Collins. 1999. Neurobiology of the structure of personality: Dopamine, facilitation of incentive motivation, and extraversion. *Behavioral and Brain Sciences* 22:3, 491–517.
6 J. Hennig. 2004. Personality, serotonin, and noradrenaline. In Robert M. Stelmack (Ed.), *On the Psychobiology of Personality: Essays in Honor of Marvin Zuckerman*, New York, Pergamon, 379–395.
7 Petra Netter. 2004. Personality and hormones. In R. M. Stelmack (Ed.), *On the psychobiology of personality: Essays in honor of Marvin Zuckerman*, New York, Elsevier, 353–377.
8 Nigel Nicholson, Emma Soane, Mark Fenton-O'Creevy, and Paul Willman. 2005. Personality and domain-specific risk taking. *Journal of Risk Research* 8:2, 157–176.
9 Suzanne Prinz, Gerhard Gründer, Ralf-Dieter Hilgers, Oliver Holtemöller, and Ingo Vernaleken. 2014. Impact of personal economic environment and personality factors on individual financial decision making. *Frontiers in Psychology* 4:5, 158.
10 Greg Filbeck, Patricia Hatfield, and Philip Horvath. 2005. Risk aversion and personality type. *Journal of Behavioral Finance* 6:4, 170–180.
11 Andreas Oehler and Florian Wedlich. 2018. The relationship of extraversion and neuroticism with risk attitude, risk perception, and return expectations. *Journal of Neuroscience, Psychology, and Economics* 11:2, 63–92.
12 Robert B. Durand, Lucia Fung, and Manapon Limkriangkrai. 2019. Myopic loss aversion, personality, and gender. *Journal of Behavioral Finance* 20:3, 339–353.
13 Robert B. Durand, Rick Newby, and Jay Sanghani. 2008. An intimate portrait of the individual investor. *Journal of Behavioral Finance* 9:4, 193–208.
14 Robert B. Durand, Rick Newby, Leila Peggs, and Michelle Siekierka. 2013. Personality. *Journal of Behavioral Finance* 14:2, 116–133.
15 Robert B. Durand, Rick Newby, Kevin Tant, and Sirimon Trepongkaruna. 2013. Overconfidence, overreaction and personality. *Review of Behavioral Finance* 5:2, 104–133.

16 Andreas Oehler, Stefan Wendt, Florian Wedlich, and Matthias Horn. 2018. Investors' personality influences investment decisions: Experimental evidence on extraversion and neuroticism. *Journal of Behavioral Finance* 19:1, 30–48.

17 Muhammad Zubair Tauni, Zia-ur-Rehman Rao, Hong-Xing Fang, and Minghao Gao. 2017. Does investor personality moderate the relationship between information sources and trading behavior? Evidence from Chinese stock market. *Managerial Finance* 43:5, 545–566.

18 Alessandro Bucciol and Luca Zarri. 2017. Do personality traits influence investors' portfolios? *Journal of Behavioral and Experimental Economics* 68:July, 1–12.

19 Ian D. Gow, Steven N. Kaplan, David F. Larcker, and Anastasia A. Zakolyukina. 2016. CEO personality and firm policies. National Bureau of Economic Research WP No. 22435.

20 Karel Hrazdil, Jiri Novak, Rafael Rogo, Christine I. Wiedman, and Ray Zhang. 2018. Measuring CEO personality using machine-learning algorithms: A study of CEO risk tolerance and audit fees. Kelley School of Business Research Paper No. 18–10.

21 Ming Liu. 2019. CEO big five personality and stock price crash risk. Available at SSRN 3312426.

22 Jochen Becker, Josip Medjedovic, and Christoph Merkle. 2019. The effect of CEO extraversion on analyst forecasts: Stereotypes and similarity bias. *Financial Review* 54:1, 133–164.

23 Sarah Brown and Karl Taylor. 2014. Household finances and the "Big Five" personality traits. *Journal of Economic Psychology* 45:December, 197–212.

24 Gianpaolo Parise and Kim Peijnenburg. 2019. Noncognitive abilities and financial distress: Evidence from a representative household panel. *Review of Financial Studies* 32:10, 3884–3919.

25 Gema Zamarro. Forthcoming. Alternative measures of non-cognitive skills and their effect on retirement preparation and financial capability. *Journal of Pension Economics & Finance*.

26 Robert D. Hare. 2003. *The Hare Psychopathy Checklist-Revised*, 2e. Toronto, ON, Multi-Health Systems.

27 Sherree DeCovny. 2012. The financial psychopath next door. *CFA Magazine* 23:2, 34–35.

28 Paul Babiak and Robert Hare. 2006. Snakes in suits: When psychopaths go to work. New York, Regan Books.

29 Laurie L. Ragatz, William Fremouw, Edward Baker. 2012. The psychological profile of white-collar offenders: Demographics, criminal thinking, psychopathic traits, and psychopathology. *Criminal Justice and Behavior* 39:7, 978–997.

30 Corey Shank. 2018. Deconstructing the corporate psychopath: An examination of deceptive behavior. *Review of Behavioral Finance* 10:2, 163–182.

31 Anna Vedel and Dorthe K. Thomsen. 2017. The dark triad across academic majors. *Personality and Individual Differences* 116:October, 86–91.

32 Christopher Marcin Kowalski, Philip A. Vernon, and Julie Aitken Schermer. 2017. Vocational interests and dark personality: Are there dark career choices? *Personality and Individual Differences* 104:January, 43–47.

33 Clinton H. Richards, Joseph Gilbert, and James R. Harris. 2002. Assessing ethics education needs in the MBA program. *Teaching Business Ethics* 6:4, 447–476.

34 M. Schneider and S. Prasso. 2002. How an MBA can bend your mind. *Bloomberg*, March 31.

35 Alain Cohn, Ernst Fehr, and Michel André Maréchal. 2014. Business culture and dishonesty in the banking industry. *Nature* 516:December, 86–89.

36 Dallas Leigh Hill. 2018. Climbing the corporate ladder: Desired skills and successful psychopaths. University of Ontario Institute of Technology master's thesis.

37 Paul Babiak, Craig S. Neumann, and Robert D. Hare. 2010. Corporate psychopathy: Talking the walk. *Behavioral Sciences and the Law* 28:2, 174–193.

38 Kent A. Kiehl, Andra M. Smith, Robert D. Hare, Adrianna Mendrek, Bruce B. Forster, Johann Brink, and Peter F. Liddle. 2001. Limbic abnormalities in affective processing by criminal psychopaths as revealed by functional magnetic resonance imaging. *Biological Psychiatry* 50:9, 677–684; Robert James R. Blair. 2007. The amygdala and ventromedial prefrontal cortex in morality and psychopathy. *Trends in Cognitive Sciences* 11:9, 387–392; Robert James R. Blair. 2008. The cognitive neuroscience of psychopathy and implications for judgments of responsibility. *Neuroethics* 1:3, 149–157; Andrea L. Glenn, Adrian Raine, and Robert A. Schug. 2009. The neural correlates of moral decision-making in psychopathy. *Molecular Psychiatry* 14:1, 5–6; Yaling Yang, Adrian Raine, Katherine L. Narr, Patrick M. Colletti, and Arthur W. Toga. 2009. Localization of deformations within the amygdala in individuals with psychopathy. *Archives of General Psychiatry* 66:9, 986–994.

39 Pamela R. Perez. 2012. The etiology of psychopathy: A neuropsychological perspective. *Aggression and Violent Behavior* 17:6, 519–522.

40 Corey A. Shank, Brice V. Dupoyet, Robert B. Durand, and Fernando Patterson. 2019. The relationship between psychopathy and financial risk and time preferences. Available at SSRN 3355774.

41 Corey A. Shank. 2018. Deconstructing the corporate psychopath: An examination of deceptive behavior. *Review of Behavioral Finance* 10:2, 163–182.

42 Charles Ham, Mark Lang, Nicholas Seybert, and Sean Wang. 2017. CFO narcissism and financial reporting quality. *Journal of Accounting Research* 55:5, 1089–1135.

43 Nihat Aktas, Eric De Bodt, Helen Bollaert, and Richard Roll. 2016. CEO narcissism and the takeover process: From private initiation to deal completion. *Journal of Financial and Quantitative Analysis* 51:1, 113–137.

44 Leanne ten Brinke, Aimee Kish, and Dacher Keltner. 2018. Hedge fund managers with psychopathic tendencies make for worse investors. *Personality and Social Psychology Bulletin* 44:2, 214–223.

45 James J. Heckman, Jora Stixrud, and Sergio Urzua. 2006. The effects of cognitive and noncognitive abilities on labor market outcomes and social behavior. *Journal of Labor Economics* 24:3, 411–482.

46 James J. Heckman and Yona Rubinstein. 2001. The importance of noncognitive skills: Lessons from the GED testing program. *American Economic Review* 91:2, 145–149.

47 B. Brau, J. Brau, A. Holmes, and M. Ringold. 2019. Factors of the desire to become a finance major: An empirical analysis. Working Paper, Brigham Young University.

48 Ulrike Malmendier and Geoffrey Tate. 2005. CEO overconfidence and corporate investment. *Journal of Finance* 60:6, 2661–2700.

49 Ulrike Malmendier and Geoffrey Tate. 2008. Who makes acquisitions? CEO overconfidence and the market's reaction. *Journal of Financial Economics* 89:1, 20–43.

50 Jeong-Bon Kim, Zheng Wang, and Liandong Zhang. 2016. CEO overconfidence and stock price crash risk. *Contemporary Accounting Research* 33:4, 1720–1749.

51 David Hirshleifer, Ankie Low, and Siew Hong Teoh. 2012. Are overconfident CEOs better innovators? *Journal of Finance* 67:4, 1457–1498.

52 J.B. Heaton. 2002. Managerial optimism and corporate finance. *Financial Management* 31:2, 33–45.

53 Camelia M. Kuhnen and Brain T. Melzer. 2018. Noncognitive abilities and financial delinquency: The role of self-efficacy in avoiding financial distress. *Journal of Finance* 73:6, 2837–2869.

54 Brain C. Cadena and Benjamin Keys. 2013. Can self-control explain avoiding free money? Evidence from interest-free student loans. *Review of Economics and Statistics* 95:4, 1117–1129.

55 Janice W. Jadlow and John C. Mowen. 2010. Comparing the traits of stock market investors and gamblers. *Journal of Behavioral Finance* 11:2, 67–81.

12 Types of Intelligence and Investment Performance

The billionaire investor Warren Buffet observes that having a superior level of intelligence, an intelligence quotient (IQ) between 120 and 140, is important for investing. However, you do not need to be a genius with an IQ above 140. Being smart is good enough. Buffet says, "You don't need to be a rocket scientist. Investing is not a game where the guy with the 160 IQ beats the guy with 130 IQ."[1] Making sound investment decisions is complex because it involves evaluating risk, uncertainty, expected returns, assets correlations and then matching personal risk aversion and financial goals. The modern world increases the difficulty through the complexity of financial instruments and the interconnectedness of global markets. Thus, it seems that intelligence should be an essential attribute for a good financial decision maker. Buffet's statement uses the best-known measure of cognitive ability—the IQ. However, there are other ways to measure intelligence. Each captures some aspect of the complexities of cognitive ability. How do the different types of cognitive ability impact investing and other economic decision making? This chapter explores these topics.

Types of Cognitive Processing

What is intelligence? People generally think of intelligence as the capacity to learn and apply knowledge and skills. However, cognitive ability may be hard to identify. For example, consider the person with little formal education but who is "street smart" compared to the physics genius next door who seems baffled by things that are common sense to others. The ambiguity may be resolved by understanding that there are different kinds of cognitive ability. Scholars have identified three distinct and independent kinds of intelligence: fluid intelligence; cognitive reflection; and the theory of mind. Strong ability in one aspect of intelligence does not necessarily mean strong ability in others.

- *Fluid intelligence* is the traditional measure of intelligence that is commonly measured by the IQ. The IQ is a gauge of a person's reasoning ability scaled to an average of 100 for the whole population. People with higher IQs may be suited to handling the complexity associated with finance and investing.

- *Cognitive reflection* is the tendency for a person to put in more time and effort than others when assessing a situation. It is an analytical mode of thinking. Intuition is at the other end of the reasoning scale from cognitive reflection. Intuition often uses heuristics and psychological biases to facilitate quicker decisions. A person who engages in more analytical reasoning is more likely to avoid behavioral biases and make more rational decisions.

- The *theory of mind* refers to a person's ability to infer the intentions of others. People have different access to information, wisdom, and analytical skills. Since the investment world involves many people (chief executive officers, analysts, brokers, portfolio managers, investors, etc.), being able to observe other people's actions and discussions can be very valuable for inferring what they know. Some people call this being "street smart."

Cognitive Processing and Investing

IQ and Investing Decisions

Are people with higher fluid intelligence more likely to be both involved in the stock market and be more successful investors? The answer appears to be "yes." Consider this specific example from Finland involving the link between IQ and investment success. Finland has a mandatory military service for young men. As a part of that service, all Finnish men complete an IQ test. Grinblatt and colleagues merged the IQ data with tax returns of approximately 160,000 Finnish men to examine the relationship between intelligence and stock market participation.[2]

Their study sorted men into nine IQ categories. The categories follow stanine groups such that category 1 (category 9) reports the lowest (highest) 4 percent of IQ in the population. Categories 2 and 8 each represent 7 percent of the population. Categories 3 and 7 represent 12 percent each of the population, categories 4 and 6 represent 17 percent each, and category 5 represents 20 percent. The distribution in the study closely followed this proportion. The results showed a positive relation between IQ scores and stock market participation. As shown in Figure 12.1, 42.5 percent of those in the second highest IQ category participated in the market, while the rate was only 13.1 percent for the second lowest IQ category. Statistical analysis showed that IQ accounts for most of the differences in the stock market participation rate, rather than income, wealth, or other characteristics. In fact, analysis of affluent Finnish men and of less affluent Finnish men showed that the role of the IQ in stock market participation was roughly the same in each group. Thus, the IQ drove stock market participation more than wealth. Additionally, men with a high IQ chose superior portfolios that were more diversified. Because the IQ test measures different cognitive

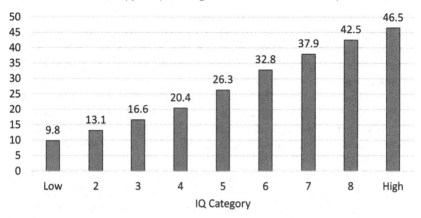

Figure 12.1 IQ and Stock Market Participation

abilities, an interesting observation was that while logical, verbal, and mathematical skills all played a role, the mathematical component was especially influential.

In another study, Grinblatt and colleagues merged the data on the IQ of Finnish men with stock trading and portfolio data.[3] Because everyone in this group had invested, the question was whether IQ made a difference in investment returns. It did. The results showed that the high IQ investors' portfolios outperformed the low IQ investors' portfolios by 4.9 percentage points per year. This higher return stems from the higher IQ investors exhibiting better market timing and stock picking skills as well as more effective tax loss selling than low IQ investors. In other words, higher IQ investors were better at buying low and selling high (market timing), buying better performing stocks (stock picking skills), and selling stocks with losses at the end of the year to lower their taxes (tax loss selling). Lower IQ investors were more prone to the disposition effect, which is the tendency to hold onto losers for too long and to sell winners too soon suggesting an inverse relationship between investment biases and intelligence.

The Finnish data was also merged with mutual fund choices.[4] The magnitude of the fees charged by mutual funds is a strong predictor of mutual fund return performance. On average, low-fee funds earn higher returns than high-fee funds. The fees that a mutual fund investor pays depend on several factors. First, actively managed funds charge higher fees than passively managed index funds. Second, funds that focus on small firm equities or emerging market equities and thus belong to niche sectors charge higher fees than domestic large equity funds. Finally, the manner in which the fund is purchased matters. Buying directly from the mutual fund is cheaper than buying through expensive retail networks that include a sales load. This study showed that high IQ investors paid lower mutual fund fees. People with a high IQ avoided high-fee funds in several ways:

- They avoided actively managed funds.
- They avoided high cost categories of funds. For example, instead of buying a high-fee balanced fund that invests in both stocks and bonds, they created a "homemade" balanced fund by buying one low-cost stock fund and one low-cost bond fund.
- Purchases were more likely to be made directly with the fund rather than through expensive retail networks. Low IQ investors were more likely to invest through retail networks because they needed more advice and service, which are associated with this more costly distribution channel.
- Even within mutual fund categories or investment philosophies, high IQ investors bought the low-fee fund. For example, among actively managed large company equity funds, high IQ investors tended to choose funds with the lowest management fees.

Inflation expectations play an important function in driving peoples' consumption, saving, and borrowing decisions. For example, if you predict high inflation in the future, you might change the makeup of your portfolio and consume more goods now while you have higher purchasing power. The Finish data was combined with the Consumer Survey of Statistics Finland to assess what role cognitive ability plays in inflation forecasting. Every month, the survey asked a repeated sample of approximately 1,500 Finns about general and personal economic conditions, including inflation expectations. The survey data covered the period from 2001 to 2015. By comparing the predicted 12-month inflation rate to the realized inflation rate, they computed the forecast error. D'Acunto and colleagues examined many aspects of this predicted inflation in relation to fluid intelligence.[5] For example, consider the value of the prediction itself. Was the value reasonable? Did it indicate a common bias by taking the form of a multiple of 5 (e.g., 2, 2.5, 3, etc.)? How significant was the forecast error? Men with an IQ above the median were called high IQ men. The study found that high IQ men displayed 50 percent lower forecast errors than low IQ men. Specifically, the average absolute forecast error for people in the smallest decile group of IQs was 4.3 percent, compared to just 0.6 percent for the largest decile group of IQs. Also, low IQ men were more likely to round and to forecast implausible values. Finally, high IQ men were the only ones to respond that they would increase their propensity to consume when expecting higher inflation.

Fluid intelligence and investing have also been explored in six European countries using the Survey of Health, Ageing, and Retirement in Europe (SHARE).[6] Christelis and colleagues' survey included personal interviews with 19,548 households and 32,405 individuals aged 50 years and over in six countries on a wide range of topics from health to wealth. The interviews contained questions related to IQ generally, and specifically numeracy, executive function, and memory. Executive function skills are the self-management skills that help people to manage their emotions, organize work, and adapt as their circumstances change. The study provided strong evidence that fluid intelligence was an essential driver for participating in the stock market. That is, higher intelligence was associated with both directly owning stocks and indirectly owning them through mutual funds. The result is not due to educational levels or age, because it held for people with and without a college education and for those both younger and older than 65 years of age.

Finally, Dohmen and colleagues engaged 952 German adults to examine whether the key traits of risk aversion and impatience were systematically related to cognitive ability.[7] The measure of cognitive ability had two aspects, verbal and nonverbal tasks, modeled after a common intelligence test. The study used paid experiments to measure the willingness to take risks and impatience. The willingness to take risks was estimated using a series of 20 choices between a paid lottery and a safe payment. For example, participants had to decide between a gamble that could win them either 300 euros or 0 euros, each with a 50 percent probability, or simply take 0 euros. The lottery gamble was always the same, but the safe payment steadily increased with each question in increments of 10 euros up to 190 euros in the twentieth question. A participant could take the gamble in the first question and continue taking the gamble until switching to the safe option. The switch to the safe option indicated their level of risk aversion. The level of participants' impatience was also measured. Participants were shown a series of questions that included a choice between accepting 100 euros now versus X euros in 12 months. For the first question, X equaled 100. Thus, all the participants accepted the 100 euros immediately rather than wait one year. Each successive question increased X by a compounded 2.5 percent. Eventually, participants chose to wait 12 months for the larger payment, which revealed their level of impatience. They found that people with lower fluid intelligence were significantly more risk averse and significantly more impatient than people with high intelligence. Their findings were robust in controlling for personal characteristics, educational attainment, and income.

What do these studies indicate about the relationship between IQ and investing? Relative to people with a low IQ, people with a higher IQ:

- were more likely to invest in the stock market;
- were better at timing the market;
- picked better performing stocks;
- formed better portfolios;
- engaged in wealth enhancing tax-loss selling;
- made better inflation forecasts;
- exhibited lower risk aversion; and
- were more financially patient.

Analytical versus Intuitive Thinking and Investing Decisions

Nobel Laureate Daniel Kahneman describes two different modes of cognitive reasoning. He outlined these modes in his Nobel lecture delivered in

Stockholm when he received the 2002 Bank of Sweden Prize in Economic Sciences. He further explored these thinking processes in his 2011 book, *Thinking, Fast and Slow*.[8] 'Thinking slow' refers to the analytical thinking mode or what he calls reasoning. It occurs when computing multiplication, such as 13 times 157. 'Thinking fast' refers to the intuitive thinking mode. Intuitive thinking relies on feelings, heuristics, and relating past experiences to form opinions and make decisions. While analytical thought is deliberate and takes effort (slow), the intuitive mode of thinking is spontaneous and effortless (fast). Everyone can think in either mode, but most people tend to default to one mode more than the other. People make most judgments and choices intuitively because it is effortless.

Consider the example of driving a car with passengers. Much of the time, driving requires only intuitive cognitive processing. During these times, the driver can drive and engage in small talk. When either the driving or the conversation becomes more complex, the driver needs to use more cognitive capacity. So, having a debate about climate change or attempting to parallel park puts the driver into an analytical thinking mode. If the driver is debating, then he becomes a distracted driver. Alternatively, if the driver is parallel parking, then the conversation becomes interrupted. Financial decisions require more cognitive capacity because they often require dealing with the complexity of uncertainty, diversification, asset allocation, market efficiency, and risk. Thus, these decisions are better suited to the analytical thinking mode. Alternatively, intuitive thinkers use more heuristics to facilitate faster thinking. Unfortunately, using heuristics is associated with behavioral biases that lead to poor financial decisions.

Measuring Analytical and Intuitive Tendencies: Cognitive Reflection

Cognitive reflection is the ability to take time to reflect a little longer before acting. Shane Frederick designed a simple three-question test to measure the extent to which a person tends toward the analytical thinking mode.[9] These questions are called the cognitive reflection test (CRT). The CRT uses questions that immediately tempt an impulsive wrong answer. To get the right answer, you need more cognitive processing. Intuitive thinking leans toward giving a quick, impulsive answer, which will be an incorrect answer in the CRT. Taking the time to switch to an analytical thinking mode is more likely to result in the correct answer. The CRT provides a range of scores between zero and three correct answers. Achieving three correct answers indicates strong analytical thinkers, while producing zero correct answers denotes strongly intuitive thinking. The questions are:

1   If it takes 5 machines 5 minutes to make 5 widgets, how long would it take 100 machines to make 100 widgets?
2   In a lake, there is a patch of lily pads. Every day, the patch doubles in size. If it takes 48 days for the patch to cover the entire lake, how long would it take for the patch to cover half the lake?

3   A bat and ball together cost $1.10. The bat costs $1.00 more than the ball. How much does the ball cost?

Notice the rhythm in the first question. The symmetry of the 5, 5, and 5 followed by the 100, 100, leads to the impulsive answer of 100. However, this is not the correct answer, which is five minutes. In the second question, the terms 'doubles' and 'half' lead the reader to half the 48 days to answer 24 days. This impulsive answer is not correct. The lily pads must have covered half the lake on day 47 in order to double and cover the full lake on the forty-eighth day. Finally, the shown costs of $1.10 and $1.00 elicit the impulsive answer of $0.10. However, if that were the case, the difference in price would be only $0.90. The ball must cost $0.05 and the bat $1.05 to have a $1.00 difference. The number of correct answers scores the CRT. A CRT score of zero or one suggests that the person is an intuitive thinker, while a CRT score of two or three classifies the person as an analytical thinker.

The original three-question CRT has become too well known and thus might have lost its usefulness. Four additional questions have been designed that can be used on their own or with the original three. The more recent four questions are:[10]

4   If John can drink one barrel of water in 6 days, and Mary can drink one barrel of water in 12 days, how long would it take them to drink one barrel of water together?
5   Jerry received both the 15th highest and the 15th lowest mark in the class. How many students are in the class?
6   A man buys a pig for $60, sells it for $70, buys it back for $80, and sells it finally for $90. How much has he made?
7   Simon decided to invest $8,000 in the stock market one day early in 2008. Six months after he invested, on July 17, the stocks he had purchased were down 50%. Fortunately for Simon, from July 17 to October 17, the stocks he had purchased went up 75%. At this point, Simon

a   has broken even in the stock market;
b   is ahead of where he began;
c   has lost money.

The common responses are (4) intuitive 9, analytical/correct 4; (5) intuitive 30, analytical/correct 29; (6) intuitive $10, analytical/correct $20; and (7) intuitive B, analytical/correct C.

Cognitive Reflection and Behavioral Biases

Is the intuitive/analytical thinking mode spectrum related to investment behavioral biases? Since intuitive thinking uses more heuristics than the analytical thinking mode, it is more susceptible to the influence of psychological biases. This is illustrated by Hoppe and Kusterer's survey of 414 university students that measured cognitive reflection as well as their susceptibility to the base rate fallacy and conservatism bias.[11] A person has engaged in the base rate fallacy when too little weight is placed on the original or base rate of possibility, and too much weight is placed on recent or transitory occurrences. Conservatism bias occurs when a person clings to his prior opinions or forecasts and disregards new information.

Consider the base rate test that uses this question:

> In a city with 100 criminals and 100,000 innocent citizens, a surveillance camera exists with automatic face recognition software. If the camera sees a known criminal, it triggers the alarm with 99 percent probability; if the camera sees an innocent citizen, it triggers the alarm with a probability of 1 percent. What is the probability of filming a criminal when the alarm is triggered?

The most common responses are probabilities greater than 90 percent, which represent a quick and intuitive answer. People focus on the 99 percent hit rate probability of the small population (criminals) without fully taking into account the small probability of a much larger population (innocents). A false positive, or false alarm, mistakenly triggers the alarm for an innocent citizen. The number of people who could trigger the false positive is far greater than the number of actual criminals. The false positives are 1 percent of 100,000, which is 1,000. There are only 100 criminals. The surveillance system will identify 99 percent of the 100 criminals, or 99 of them. So, if everyone walked by the camera, it would trigger the alarm 1,099 times. Only 99 of triggers would be criminals. The correct 99 of 1,099 represents only 9 percent, which is the answer. That answer is computed as (= [0.99 x 100] ÷ [0.99 x 100 + 0.01 x 100,000]). Getting the correct answer, or even a good estimation, requires a reflective cognitive process. Indeed, for the study, the CRT scores were higher for those who responded that the probability was 10 percent or less.

A similar behavioral bias is conservatism bias, which is a failure to adapt prior beliefs to new evidence. In this question, researchers told participants about two urns (A and B). Urn A contained seven red balls and three blue balls. Urn B contained three red balls and seven blue balls. The participants drew 12 balls from an unlabeled urn with replacement, which means that after seeing the ball, the ball is put back into the urn. The question was, "What is the probability that the urn was urn A when observing the result of eight red balls and four blue balls?"

Here the base rate is 50 percent because if the color of the 12 randomly drawn balls was not given, there was a 50–50 chance they came from urn A or urn B. However, the distribution of the colors suggests that the balls were likely drawn from urn A because it has a higher proportion of red balls, increasing the probability that a red ball would be pulled out. However, what is the probability? The answer is 97 percent. The average CRT score for those who provided an answer of more than 89 percent was higher than the CRT scores of people who answered less than 90 percent. The most common answer was the base rate of 50 percent. Those participants failed to update the base rate with the new information of balls drawn. They also had lower CRT scores.

The previous study surveyed students. How are better-trained and more-experienced investors likely to behave? Nofsinger and Varma tested 108 financial planners to determine their thinking mode and its relation to risk aversion and behavioral biases.[12] After taking the CRT, researchers grouped the advisors' answers into intuitive (n = 43) and analytical (n = 65) thinkers for the analysis. Additional questions tested their risk aversion, patience, and prospect theory behavior. First, consider two questions that focused on the principles of prospect theory. Prospect theory states that people tend to choose the certain option when it is framed in a positive manner and choose the risky option when it is framed in a negative way.

- Which investment payoff would you pick? Receive (A) $100 for certain or (B) a 50 percent chance to receive $300 and a 50 percent chance to receive nothing.
- Which investment payoff would you pick? Lose (A) $100 for certain or (B) a 50 percent chance to pay $300 and a 50 percent chance to pay nothing.

The first question used a positive frame domain of gaining money. Note that the certain payoff of $100 was less than the expected value of the gamble, $150 (= 0.5($300) + 0.5($0)). The difference of $50 might be considered a risk premium, which is the reward for taking risk. The second question used a negative frame of losing money. Again, the certain alternative had a higher expected value (−$100) than the gamble (−$150). For the first question in the positive domain, 63.4 percent of the intuitive financial planners selected the certain choice (A). However, in the loss domain questions, the majority (60.5 percent) selected the risky option (B). Although this behavior is consistent with prospect theory, the analytical planners tended toward the opposite behavior. They selected the risky option (57.9 percent) in the positive domain and the certain option (57.9 percent) in the negative domain. Both results were important behavioral finance findings. First, all financial advisors did not have the same level of risk aversion. Instead, they were risk seeking in one situation and risk averse in another. Second, intuitive thinkers were more likely to behave along the axioms of prospect theory than analytical thinkers were.

This study also examined other aspects of how cognitive reflection may affect planners in financial decision making. For example, when asked whether they preferred to receive $3,400 three months from now or $3,800 six months from now, 68.4 percent of the analytical planners recognized that this intertemporal choice problem represented a high rate of return to wait (i.e., a compounded annualized return of 56 percent) and selected the delayed payment. Intertemporal choice refers to the decisions a person makes at different points in time. In other words, this economic term describes how a person's current decisions affect what options become available in the future. Fewer than half of the intuitive planners selected the delayed $3,800. Also, when asked how much money they would spend to receive a book they ordered sooner, the average intuitive planner answered $10.03, whereas the average analytical planner answered only $2.63. The response to both questions illustrated that analytical people are more patient than intuitive people. Oechssler and colleagues also found in a sample of mostly German students that intuitive thinkers were more risk averse and impatient than analytical thinkers.[13]

A third study compared professional traders, financially oriented bank employees, and non-financial people. Thoma and colleagues refer to the analytical and intuitive framework as good thinking versus gut feeling.[14] The survey respondents consisted of 44 traders from U.K. investment banks and trading houses, 53 banking finance workers, and 57 non-banking workers. The traders scored the highest on the three-question CRT test, averaging 2.7 (out of a total of three) compared to 1.6 for the finance workers and 1.33 for the non-bank people. This result might be surprising to some who think that trading is about having an intuition about market dynamics. Yet experienced traders scored highly in cognitive reflection. To some extent, this result may reflect a survival bias. That is, traders may need to be cognitively reflective to survive in that business over time. The study also showed that traders were willing to take more financial risk than bankers and people with non-finance backgrounds. Bank employees with financial training were willing to take more financial and non-financial risk than individuals working elsewhere. Overall, the evidence suggests that people with lower cognitive reflection (intuitive thinkers) are less likely to take financial risk.

The Theory of Mind and Investing Decisions

Texas Hold'em poker has become popular over the past decade. This form of poker, like all others, requires intelligence to be successful. A high level of fluid intelligence is helpful for understanding the probabilities of winning for various cards held each time cards are revealed. A complication is that poker is played against other people. The elite players will excel at inferring what cards their opponents hold through their bets and behavior. "Reading," or assessing, your opponent uses the theory of mind cognitive process. The theory of mind cognition is the capacity to infer other people's intentions.

To determine whether people use the theory of mind cognitive process while participating in financial markets, Bruguier and colleagues conducted an experiment in which participants traded with one another.[15] Each person started with various amounts of cash at the beginning of the trading game, and two risky assets that could pay a dividend between $0 and $0.50 each. The dividends of the two risky assets had a negative correlation, which is a relation between two variables in which one variable increases as the other decreases, and vice versa. However, the researchers gave only some people inside information about the dividend of one of the risky assets. For example, a participant knew that the dividend of risky asset B would be above $0.35. Thus, asset B would be more valuable because that meant that asset A would have a low dividend. The trader with inside information would buy more of asset B than asset A. The other participants did not have this information but were told that insiders existed in the market who held information they did not. If the other participants detected one trader buying a considerable amount of asset B, they could infer that the trader knew that asset B was worth more than its current price and then trade accordingly for themselves. This inference of information is theory of mind cognitive processing. Active trading of the assets commenced.

After conducting this experiment 13 times with 20 different participants each time, the researchers concluded that the people "are rather good at inferring information from the order flow, despite their lack of formal financial training." The participants without private information correctly forecasted the price direction two-thirds of the time, which illustrated that they correctly identified the informed trader and adjusted their trades accordingly.

Additionally, Bruguier and colleagues used an IQ test to measure for fluid intelligence and the CRT for cognitive reflection. They used two traditional tests for theory of mind intelligence. The first is a movie clip with geometric shapes imitating social interaction through movement. The clip is stopped every five seconds, and participants are asked to predict the next movement of the shapes. The second task was based on pictures of peoples' eye gaze. The participants were asked to pick among four mental states that the eye gazes depicted.

Interestingly, the study reported that the results of a math test, such as a subcategory of an IQ test, were not related to price forecasting success. Success in inferring the information held by other traders came from theory of mind processing, not fluid intelligence. Finally, the study also conducted brain imaging with functional magnetic resonance imaging (fMRI) and found that theory of mind originates in the most developed part of the brain, the frontal and medial parts of the cortex.

How Each Type of Intelligence Impacts Investing Decisions

Fluid intelligence, cognitive reflection (analytical/intuitive), and theory of mind all have been shown to influence investing decisions. Do they all influence the same part of the trading process, or do they have separate impacts? Conducting experiments allows researchers to focus on the factor that they think has an influence while controlling all other aspects to ensure that they are not interfering. In one experiment, Corgnet and colleagues directed 167 participants to engage in a computerized 17-round trading exercise that generated 2,839 profit outcomes.[16] In each round, every participant had the opportunity to trade an asset that could have one of three values, but they did not know what the value would be. Each had private information about the true value and could try to infer the information others had through their trades. The private information was that each trader was informed of one possible value the asset could not take.

In order to be successful, the trader would need three types of skills to value the asset: math/statistical valuation; assessment of market signals; and avoidance of behavioral biases. In other words, the trader would need to use the information available to value the asset to determine if the current price allows for an opportunity, to infer the private information held by others through their trades, and to execute rational trading decisions. The researchers measured each participant's IQ, cognitive reflection, and theory of mind and related it to trades and profits. The main findings are summarized as:

- Fluid intelligence provided traders with computational skills to model asset value.
- Cognitive reflection helped traders to avoid behavioral biases and update their prior beliefs to execute better trades.
- Theory of mind allowed traders to assess the information content of other traders' orders.
- Cognitive reflection and theory of mind were complementary because traders benefited from interpreting market signals when they can execute trades well.

Summary

There are three distinctly separate kinds of intelligence: fluid intelligence; cognitive reflection; and the theory of mind. Fluid intelligence is the traditional measure of a person's reasoning ability commonly measured by the IQ. Cognitive reflection is the tendency to take more time and effort (analytical mode) when assessing a situation rather than spontaneous and effortless decision making (intuitive mode). The theory of mind refers to a person's ability to infer the intentions of others.

Investors with a higher IQ are more likely to invest in the stock market and to have the computational skills to buy low and sell high (market timing), buy better performing stocks (stock picking skills), and sell stocks with losses at the end of the year to lower their taxes (tax loss selling). Investors with a lower IQ are more prone to behavioral biases like the disposition effect.

An analytical thinking mode is better suited to dealing with the complexity of uncertainty, diversification, asset allocation, market efficiency, and risk. The intuitive thinking mode uses more heuristics to facilitate faster thinking, which is associated with behavioral biases. The CRT measures the degree of analytical versus intuitive thinking. People with a high CRT score (analytical thinkers) are more willing to take financial risk and are more able to execute better trades. People with a low CRT score (intuitive thinkers) use a higher intertemporal discount rate and are more susceptible to the axioms of prospect theory, base rate fallacy, and conservatism bias.

The theory of mind cognitive process is the capacity to infer other people's intentions. People with high theory of mind cognition are more able to assess the information content of other traders' orders and make better investment predictions. Cognitive reflection and theory of mind are complementary because traders benefit when they can execute trades well after interpreting market signals.

Questions

1   What are the three different measures of intelligence?
2   How does fluid intelligence impact financial decisions?
3   How does cognitive reflection relate to financial decisions?
4   What is the relationship between the theory of mind and investment decisions?

Notes

1   Carol Loomis. 2012. *Tap Dancing to Work: Warren Buffett on Practically Everything, 1966–2012*. New York, Penguin.
2   Mark Grinblatt, Matti Keloharju, and Juhani T. Linnainmaa. 2011. IQ and stock market participation. *Journal of Finance* 66:6, 2121–2164.
3   Mark Grinblatt, Matti Keloharju, and Juhani T. Linnainmaa. 2012. IQ, trading behavior, and performance. *Journal of Financial Economics* 104:2, 339–362.
4   Mark Grinblatt, Seppo Ikäheimo, Matti Keloharu, and Samuli Knüpfer, 2016. IQ and mutual fund choice. *Management Science* 62:4, 924–944.
5   Francesco D'Acunto, Daniel Hoang, Maritta Paloviita, and Michael Weber. 2019. IQ, expectations, and choice. National Bureau of Economics Research, NBER No. 25496.
6   Dimitris Christelis, Tullio Jappelli, and Mario Padula. 2010. Cognitive abilities and portfolio choice. *European Economic Review* 54:1, 18–38.
7   Thomas Dohmen, Armin Falk, David Huffman, and Uwe Sunde. 2010. Are risk aversion and impatience related to cognitive ability? *American Economic Review* 100:3, 1238–1260.
8   Daniel Kahneman. 2011. *Thinking, fast and slow*. New York, Farrar, Straus and Giroux.
9   Shane Frederick. 2005. Cognitive reflection and decision making. *Journal of Economic Perspectives* 19:4, 25–42.
10 Maggie E. Toplak, Richard F. West, and Keith E. Stanovich. 2014. Assessing miserly information processing: An expansion of the cognitive reflection test. *Thinking & Reasoning* 20:2, 147–168.
11 Eva I. Hoppe and David J. Kusterer. 2011. Behavioral biases and cognitive reflection. *Economics Letters* 110:2, 97–100.
12 John Nofsinger and Abhishek Varma. 2011. How analytical is your financial advisor? *Financial Services Review* 16:4, 245–260.
13 Jörg Oechssler, Andreas Roider, and Patrick W. Schmitz. 2009. Cognitive abilities and behavioral biases. *Journal of Economic Behavior & Organization* 72, 147–152.

14 Volker Thoma, Elliott White, Asha Panigrahi, Vanessa Strowger, and Irina Anderson. 2015. Good thinking or gut feeling? Cognitive reflection and intuition in traders, bankers and financial non-experts. *PloS ONE* 10:4, 1–17.
15 Antoine J. Bruguier, Steven R. Quarz, and Peter Bossaerts. 2010. Exploring the nature of "trader intuition." *Journal of Finance* 65:5, 1703–1723.
16 Brice Corgnet, Mark DeSantis, and David Porter. 2018. What makes a good trader? On the role of intuition and reflection on trader performance. *Journal of Finance* 73:3, 1113–1137.

13 Impaired Cognitive Function and Diminished Decision Making

The importance of studying cognitive aging and financial decision making has never been greater. Much of the developed world is experiencing an aging population. For example, in the United States, around 10,000 people turn 65 every day. These are the people belonging to the baby boom generation, who were born between 1946 and 1964. Additionally, European demographics are similar to that of the United States. Aging in the Japanese population is even further along, as there is a higher proportion of seniors in the population than in other countries. This generation represents a large portion of society's wealth demonstrating the importance of examining cognitive aging and financial decisions. If cognitive ability changes as one enters the senior years and if aging impacts investment decisions these factors may have major negative consequences for wealth and income. Additionally, investment decisions have become more important because of society's transition from defined benefit retirement plans to defined contribution retirements. Also, investment decisions are now made over more extended periods as lifespans have progressively gotten longer.[1] Given the size and wealth of the "boomer" generation, its investing behavior may have implications for the stock market.

Aging and Cognitive Ability

Some cognitive functions decline over time as a normal aspect of aging. Specifically, cognitive aging refers to the gradual decline over time of memory, conceptual reasoning, and processing speed. One ramification of cognitive aging is the decline in financial literacy, which is a current issue for all generations. Although seniors are generally aware of this problem, they do not seem to relate their decline in cognitive ability to their capability to manage their own finances.[2] Research estimates that cognitive ability peaks at around 20 years of age and then declines by 1 percent per year through to the age of 80, which suggests that a person loses 60 percent of their mental ability over a lifetime.[3] However, the ability to make sound financial decisions does not directly follow cognitive ability because of increases in financial knowledge and experience. Young people may have the highest cognitive ability, but they lack experience. Gaining investment knowledge and experience dramatically reduces the rate of financial errors. Experience improves financial decision making and peaks in the early fifties. This peak represents a period of maximum experience while still having good cognitive ability. Because cognitive aging continues, and little more experience is gained, the incidence of financial mistakes increases again after this peak. This pattern is known as the U-shape in financial mistakes over time and can be seen in Figure 13.1. Note the acceleration in the propensity for financial mistakes after the age of 70.

Decision Processing and Aging

Researchers have been able to study the subcortical responses during antici- pation of uncertain gains and losses throughout the age spectrum using functional magnetic resonance imaging (fMRI) and other tools during monetary incentive tasks. When anticipating monetary gains, these studies suggest that older and younger adults show similar subcortical ventral striatal responses[4] and similar neural responses to reward outcomes in the medial prefrontal cortex and ventral striatum.[5] A very different pattern between young and old adults emerges for the anticipation of monetary losses. Compared to younger adults, older adults report lower levels of anticipatory negative arousal when anticipating losses. Associated cognitive processing while anticipating a monetary loss shows reduced reactivity in the dorsal striatum and anterior insula in older adults compared to younger adults.[6] This age-related asymmetry in anticipating gains and losses is con- sistent with research that demonstrates that emotional experience in everyday life becomes less negative across adulthood.[7]

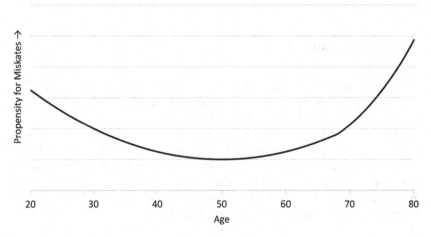

Figure 13.1 Financial Cognitive Ability and Age

A number of studies have investigated financial risk aversion and age using fMRI and similar equipment. Some studies concluded that older adults were more risk averse than younger adults. Yet some studies found no difference in risk aversion. The difference in findings appears to be caused by whether the risky tasks in the experiments required learning. Learning tasks seemed to have impacted older adults, causing them to select less risky choices due to having lower cognitive function to learn the task.[8] Indeed, neuroimaging studies that do not require rapid learning found similar patterns of neural activity in striatal[9] and prefrontal[10] regions between younger and older adults. It appears that the processing differences between older and younger adults during learning tasks stems from the connections from the medial prefrontal cortex to the ventral striatum. Age-related degradation of these connections might influence reward learning and risk taking. Researchers can now visualize the integrity of subcortical tracts using diffusion tensor imaging tractography. Samanez-Larkin and colleagues examined the structural integrity of reward circuitry in a group of younger, middle-aged, and older adults who also completed a reward learning task.[11] The study found the reduced integrity of pathways connecting (1) the thalamus to the prefrontal cortex and (2) the prefrontal cortex to the ventral striatum in older adults. This makes cognitive processing longer and increases cognitive effort. The reduced integrity was also associated with degradation in reward learning. Thus, age differences in prefrontal networks associated with increased task demands were the result of cognitive deficits.

It is also important to point out that older adults did not underperform against younger adults in all decision-making scenarios. Indeed, older adults outperformed in scenarios involving decisions that did not require learning in a new environment but instead required an accumulation of experience or emotional and motivational processing.

Aging and Financial Literacy

Financial Knowledge

Finke and colleagues directly measured the decline in financial literacy in 3,873 people over the age of 60 using the Consumer Finance Monthly (CFM) survey.[12] The CFM used the 16 questions of the Financial Literacy Assessment Test, which asked four questions in each of the categories of basic, credit, investing, and insurance. An example of an investment question is: "A young investor willing to take moderate risk for above average growth would be most interested in? 1. Treasury bills, 2. Money market mutual funds, 3. Balanced stock fund." Figure 13.2 shows the severe decline in financial literacy associated with aging. In the 60–69 age group, the average financial literacy score was 62 percent correct. That score fell to 49 percent for the 70–79 group and fell to nearly half for the 80 and over seniors. The investment literacy scores fell more than half from 57 percent correct in the 60–69 age group to 25 percent in the

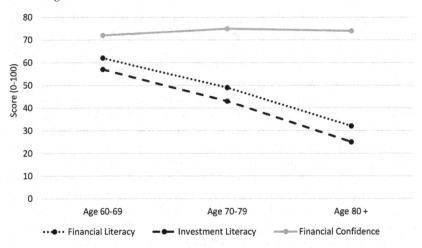

Figure 13.2 Financial Knowledge and Confidence

over-80 age group. Possibly of greater concern was that when asked how confident they were in managing their finances, the responses indicated that confidence increased from 72 percent out of 100 percent to 74 percent. Thus, older seniors failed to realize how their decline could impact their ability to manage their own finances.

Is this decline in financial literacy associated with cognitive aging? The CFM survey did not include cognitive ability questions. So, Finke and colleagues conducted further tests using the Health and Retirement Study (HRS) survey which is a national panel managed by the Institute for Social Research at the University of Michigan in cooperation with the National Institute on Aging. This survey included intelligence evaluation questions and some financial literacy questions. Using five-year age groups starting at the age of 60, the study concluded that the change in cognitive ability explained much of the decline in financial literacy. Lusardi and colleagues also used the HRS survey to explore the level of financial sophistication in subgroups of older Americans.[13] In addition to examining financial sophistication, they also varied the wording on the financial literacy questions to examine whether framing impacts responses. Finally, they examined the propensity to give "I don't know" answers to financial questions. The study found that the least educated, women, and those aged 75 and over were significantly less sophisticated about financial matters. Women were particularly sensitive to the wording of questions when financial terminology was used, thus suffering from a framing bias.

The above studies tested for changes in cognitive skills and financial literacy by examining groups of people at different ages at a specific point in time. This could be problematic as the findings could have been driven by generational differences in education and experiences rather than changes due to age. Therefore, Gamble and colleagues examined the changes in the same people over time.[14] The team used a dataset collected by the Rush Memory and Aging Project, which is an ongoing longitudinal study of aging that began in 1997 and has more than 1,500 participants from the Chicago metropolitan area on the roll. Each year, participants were given risk factor assessments and clinical evaluations, which included cognitive tests. Not every participant completed all the same surveys and interviews, and some participants have died. For their study, 575 people without dementia completed at least two of the annual evaluations. The average age was 82.2 years, and 77 percent of them were female. The scholars matched the change in cognition scores with the concurrent change in financial literacy scores. They reported the following findings:

- Two-thirds of the participants experienced cognitive decline. It was a modest change but measured over only a two- or three-year time period.
- The participants experienced a modest decline in financial literacy over a short period.
- The declines in cognition and financial literacy was positively correlated, with the strongest results in semantic memory (general world knowledge) cognition tests.
- A decrease in cognition was associated with a drop in self-confidence in general, but not a decline in confidence in managing one's own finances nor a decline in confidence in one's financial knowledge.
- While a decline in cognitive ability increased the likelihood that they got financial help, 88 percent of the participants were primarily or jointly with a spouse responsible for their financial decisions, and just 25 percent got help from someone other than a spouse.

Seeking Financial Advice

Given that cognitive ability and financial literacy decline after the age of 60, it might be surprising that studies show that only one-quarter to one-third of seniors seek financial advice. To learn more about this phenomenon, Kim and colleagues included a financial advice module in the 2016 HRS survey.[15] In addition to the regular questions asked in the HRS, the new module asked people aged 50 and over questions about whether they had obtained financial advice. Those who had not were asked why not. The survey resulted in 1,180 age-eligible respondents. The scholars sought to determine whether older people experiencing cognitive aging were likely to seek financial advice. In addition, the quality of financial advice was assessed.

The researchers reported that cognitive ability and financial literacy did not affect the likelihood of seeking financial advice. That is, people of all cognitive abilities seemed to have the same chance of seeking financial advice. However, cognitive ability and financial literacy did influence the quality of financial advice received. Specifically, people with higher cognitive ability were more likely to seek advice about more sophisticated financial tasks such as investment topics rather than how to pay the bills. The financially literate and higher cognitive ability people were also more likely to seek financial advice from professional advisors rather than "free" financial consultations that likely had conflicts of interest. For those who had not sought financial advice, higher cognitive ability people responded that they lacked trust in financial advisors.

Credit Choices

One way to examine a person's financial literacy is to view their loan choices. For example, for people with below-average credit risk, such as prime borrowers, researchers found that middle-aged people pay lower fees and borrow at a lower interest rate compared to younger and older people.[16] For home equity loans and lines of credit, the difference was as much as 0.5 percent annual percentage rate (APR). While this may not seem like much, the difference between 4.5 percent APR and 5 percent APR on a $200,000 30-year mortgage is nearly $22,000 in lifetime payments and about 11 percent of the loan value. Additionally, this result is an example of the U-shape pattern of financial mistakes over time similar to the graph in Figure 13.1.

Another example of financial knowledge is using credit card balance transfer offers. As part of their marketing campaigns, credit card companies offer to transfer the balance from one of your existing credit cards to a new card issued by them. The offer comes with a teaser rate, which is a very low interest rate (sometimes even 0 percent) for a fixed time period, e.g., six months to one year. The typical teaser rate deal is only applicable to the transferred debt. New purchases on the new card are charged a higher interest rate. Monthly payments are used to pay down the transferred balance while the new purchase debt grows at the higher rate. The optimal credit card strategy is to transfer the balance from the old card to the new card but to continue to make any new purchases on the old credit card during the new card teaser rate period. The optimal strategy was recognized from the beginning by about one-third of the people transferring their balance.[17] Slightly less than one-third of the people began making purchases on the new card but quickly recognized their mistake after one or more payment cycles. The other one-third of the people transferring their balance never discovered the optimal strategy and made purchases on the new card throughout the teaser rate period. Investigating this loan behavior and age revealed that nearly half of younger people and older people never discovered the optimal strategy. Only about 30 percent of middle-aged borrowers failed to use the optimal strategy. In addition, more middle-aged people discovered the optimal strategy immediately, and if they did not, then they were more likely to discover it quickly during the teaser rate period, once again demonstrating the U-shaped pattern in financial decision making.

Instead of specific credit choices like the credit card teaser rate example, scholars might also examine a lifetime of credit choices as revealed by a person's credit score. Using a survey and credit scores of 448 adults of varied ages, Li and colleagues examined whether higher levels of financial expertise allowed older adults a route to sound financial decision making when cognitive ability was in decline.[18] They found that financial literacy was positively associated with credit scores while controlling for cognitive ability. Since financial knowledge could be attributed to financial experience, the study separately measured experience as self-reported activities on 20 different types of financial instruments (e.g., checking accounts, credit cards, mortgages, mutual funds, etc.). Controlling for financial experience, the effect of financial knowledge on credit scores remained strong. The study concluded that comprehending financial products, not just having experience of using them, can substitute for cognitive ability in seniors. This accumulated knowledge and expertise dramatically reduced the need for cognitive information processing and active search processes. One caveat was that it is more cognitively difficult to obtain this financial expertise after the age of 60 due to cognitive aging.

Aging and Financial Fraud

A person's peak level of wealth likely occurs at the time of retirement, which comes from decades of saving and investing. The combination of this wealth and cognitive aging provides opportunities for seniors to be victimized by financial scams and fraud. Seniors experience higher rates of fraud than other age groups. The Certified Financial Planner Board of Standards surveyed 2,649 of its financial advisors with the Certified Financial Planner (CFP) designation about their older clients' experiences.[19] They concluded that seniors are "subject to unfair, deceptive or abusive financial practices, and the problem is pervasive." The majority of CFP advisors had at least one client who had been subject to these abusive practices. While the median client loss was a substantial $50,000, some clients experienced extreme losses, which skewed the average to about $140,000.

Deceptive and abusive financial practices come in many forms. It comes from outsiders as well as financial professionals and family members. Outsiders hook the clients with "no-risk high-yield investment" offers, "you have won a prize" scams, as well as Ponzi schemes, affinity fraud, foreign lotteries, Veteran's Affairs benefits' fraud, and Social Security or Medicare fraud. A Ponzi scheme is a fake investment operation where the operator pays returns to the early investors using cash contributed by later investors. However, no serious investment activities occur. Eventually, the amount of new cash from new investors is not sufficient to pay any of the older investors and the scheme collapses and everybody losses most of their investment. Bernie Madoff is likely the most well-known perpetrator of this scam as his Ponzi scheme stole approximately $65 billion from some 5,000 investors. Affinity fraud is a scheme in which the operator targets a close-knit group such as a church organization, an ethnic or neighborhood society, or a professional group. If the perpetrator can deceive a few members of the group into believing the investment he offers is good, those members unwittingly spread the word about the "opportunity" to other members. People will often fail to investigate the investment when they believe that others in their trusted group already have. Common financial exploitation coming from financial professionals include the sale of variable annuities, variable life insurance, and whole life insurance, which carry very high fees. Unfortunately, as seniors age, they become more susceptible to financial abuse from guardians and family members possibly due to cognitive aging.

Cognitive Aging and Financial Fraud

Boyle and colleagues attempted to assess whether cognitive aging plays a role in financial fraud.[20] The team used a dataset collected by the Rush Memory and Aging Project, which is described above. One crucial question tracked in the project was whether in the past year the participants had become victims of financial fraud or had been told that they were. This process enabled researchers to examine whether the participants had experienced cognitive declines from the annual cognitive tests and how any decline related to experiencing financial fraud. They found that a decline in cognitive ability of 34 percent increased the odds of becoming a victim of financial fraud by 33 percent.

Does cognitive aging impact a person's susceptibility for fraud? The scholars investigated these questions by measuring susceptibility through six questions designed to assess behaviors that are associated with being vulnerable to perpetrators. The first six questions were answered using a seven-point scale that ranged from strongly disagree to strongly agree. The sixth question was simply answered "yes" or "no." The six survey questions were:

1   I answer the phone whenever it rings, even if I do not know who is calling.
2   I have difficulty ending a phone call, even if the caller is a telemarketer, someone I do not know, or someone I did not wish to call me.
3   If something sounds too good to be true, it usually is.
4   Persons over the age of 65 are often targeted by con-artists.
5   If a telemarketer calls me, I usually listen to what they have to say.
6   Are you listed on the national do not call registry? Yes or no?

The responses were combined to form a "score of susceptibility to scam" index. They concluded that declining cognitive ability increased a person's susceptibility to being scammed.

The researchers also investigated two behavioral biases: overconfidence and the propensity to take risks in order to break even.[21] They gave participants a standard financial literacy test. Overconfidence was indicated when participants stated that they performed better on the test questions than they actually did. The researchers found that overconfidence played an essential role in becoming a financial fraud victim as a 34 percent increase in overconfidence increased the odds of being scammed by 26 percent. Finally, they examined how fraud victims changed their risk after the fraud. After taking a loss, many of the people showed an increased willingness to take on risk in an effort to break even. This relationship was stronger for older financial fraud victims. Although the general sample of elderly participants increased risk aversion over the overall sample, the fraud victims slightly decreased their risk aversion, making them more susceptible to additional fraud overtures. Indeed, a common "follow-up" scam is to offer a fraud victim the opportunity to get their money back for a substantial fee.

Investment Decisions of Seniors

Aging and Financial Risk Aversion

A person's financial risk aversion increases as he nears retirement age and then continues through his retirement years. What is the cause of this change in financial risk aversion? One possibility is a change in investing needs and goals. For example, it is standard financial advice to reduce risk exposure as one nears retirement—a common rule of thumb is to have stock exposure in a portfolio with a proportion of (100 − age) percent. That is, a 60-year-old would be advised to have 40 percent of the portfolio in stocks. This comes from the fact that seniors have different investment needs than younger people. Seniors have a shorter time horizon and income needs, whereas a young person invests for capital growth to save for the future. For example, if the stock market declined by 40 percent, someone who is 25 years old would have plenty of time to make the money back over the next market cycles. However, if someone wanted to retire in three years, he would not have much time for the stock market to recover. Thus, the near retiree should not be overly invested in equities or risk delaying their retirement. Another source of changing risk aversion is changing cognitive ability. Research shows that on average people of higher cognitive ability have lower financial risk aversion (see Chapter 13). When people experience cognitive aging, their cognitive ability declines, which might result in increased financial risk aversion. Thus, changes in financial risk aversion over time could stem from changing investing needs, declining cognitive ability, or both.

To examine the differences in financial risk aversion at different ages, Kim and colleagues use the HRS survey data.[22] The sample included 6,614 people who were 65 years or older. The scholars utilized questions in the survey regarding stock ownership (directly or through mutual funds), recent purchases of stock, bequest motivation, and cognitive assessment. The inclusion of the participant's motivation to leave an inheritance adds a new dimension in that it increases the senior's portfolio time horizon. Instead of thinking that the portfolio has an expected time horizon of the participant's life expectancy, the portfolio can be considered as having a much longer time horizon. Thus, a bequest motivation provides an incentive to invest in the stock market. However, declining cognitive ability could hinder stock market participation despite bequest motivation. Another innovation is the study's investigation of recent stock purchases. Holding stock could be the result of inertia while purchasing stocks requires a recent decision. The HRS survey included questions that comprised a mental status score and a memory score using the Telephone Interview for Cognitive Status questions to measure cognitive ability. This is a set of 20 questions and tasks frequently used in the medical sciences to classify persons who may be in cognitive decline. The bequest motivation came from a question that asked about the probability (0 to 100) of the participant leaving an inheritance totaling $10,000 or more. By analyzing whether the elderly invested in the stock market in the presence of a bequest motive and cognitive aging, the scholars found that:

- About a third of the elderly participants held stocks.
- About 36 percent of those who held stocks had also purchased stocks in the past two years.
- Cognitive ability was positively associated with stock ownership.
- Cognitive ability was not a significant factor in recent stock purchases.
- With both a strong bequest motivation and poor cognitive abilities, the cognitive aging overwhelmed the bequest motivation leading to a lower likelihood of holding stocks.

Another study used the Survey of Health, Ageing, and Retirement in Europe (SHARE) to separate the role of age-related investment needs and cognitive ability. SHARE encompassed more than 12,000 people aged 50 years or older in 11 European countries. Questions were included about each person's age, stock market participation, financial risk aversion, and cognitive skills in math, memory, and verbal fluency. Bonsang and Dohmen found that an index of these three cognitive abilities (math, memory, and verbal) was a strong factor in the level of financial risk participants were willing to take and whether they invested in the stock market.[23] By controlling for a person's age, the analysis thus controls for a person's investment needs. The study reported that people with higher levels of cognitive ability were more willing to take financial risk after controlling for age. But not all measures of cognitive ability had the same individual impact on risk aversion. The math-oriented cognitive measure had the most substantial influence. Verbal fluency also had a strong influence on risk taking. However, the memory test of cognition had only a marginal association with risk aversion. Overall, the study concluded that the financial risk aversion related to aging was more related to cognitive aging than changes in financial needs.

Aging and Portfolio Construction and Performance

Portfolio choice and performance is dependent on a person's decision-making ability and risk aversion. Cognitive aging negatively affects decision making and risk tolerance. As a balance to cognitive aging, gaining investment experience can positively influence portfolio construction decisions. Korniotis and Kumar studied more than 62,000 investor accounts from a U.S. discount brokerage from 1991 to 1996 to assess which effect dominated as investors got older.[24] The data included investment holdings and trades and demographic information about age, income, wealth, etc. They measured experience using the time the brokerage account had been open. Using the portfolio holdings and trades, the study examined various portfolio characteristics such as performance, trading, and diversification in the context of investor age and experience. The portfolio performance was reported as an annualized abnormal return using a characteristics-based asset pricing model. The model reports the extra return (called an abnormal return) earned on the portfolio above or below what should have been earned given the investment style displayed.

The findings are illustrated in Figure 13.3. Note that the portfolio performance exhibited an inverted U-pattern where the peak performance takes place at around 42 years of age. It appears that experience increased performance until a person's forties, but then cognitive aging started to degrade performance. Noticeable declines in performance began in the mid-fifties. Finally, there was an abrupt decline in performance at around the age of 70. Portfolio performance comes from underlying characteristics like stock picking ability and diversification. Drilling down into these characteristics, the study found that older investors were more financially risk averse. The increased experience led to (1) better diversification; (2) less frequent trading; (3) increased propensity for smart year-end tax-loss selling; and (4) fewer behavioral biases such as the disposition effect and familiarity bias. After controlling

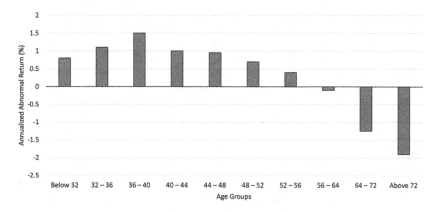

Figure 13.3 Annualized Abnormal Return by Age Group

for experience, older investors exhibited poor investment decisions, particularly in (1) diminished stock selection ability; and (2) poor diversification. The study concluded that declining cognitive ability resulted in the underperformance of 3 percentage points annually. This study illustrated the U-shape pattern of investment decision-making mistakes over the lifecycle.

Another way to examine the role of cognitive aging in portfolio decisions is to study how seniors react to an investment shock. The financial crisis of 2007 to 2008 resulted in just such a dramatic drop in the stock market. Browning and Finke used the HRS data to explore reallocations away from the stock market by retirees during the Great Recession.[25] How should an investor react to a large stock market decline? A buy-and-hold strategy would result in no reallocation. Alternatively, an investor could buy more stock to maintain asset allocation targets. A "buy low" strategy would also cause an increased allocation to stocks. An investor might sell some of the stocks with losses for tax-loss selling purposes, but then buy different stocks to maintain the portfolio allocation targets which would result in no net change in allocation to stocks. Changing the portfolio allocation targets to reduce stock exposure after a dramatic decline prevents investors from benefiting from the eventual stock market recovery. Reducing exposure to the stock market after a decline is usually irrational and the result of an emotional response. Avoiding market timing mistakes requires the regulation of emotions during the decision process. Over 57 percent of the sample decreased their allocation to the stock market beyond what was explained by the portfolio returns. People suffering from cognitive aging may be more susceptible to emotion-driven trading. The study found a negative relationship between cognitive ability and shifts away from the stock market. Cognitive decline impairs a retiree's ability to manage emotions in the investment decision process.

Aging and Behavioral Biases

People may use heuristics to make decisions when faced with complex decisions. Using these simple rules reduces the amount of cognitive processing needed. As cognitive aging progresses, people may employ more types of heuristics and do so more often. To investigate this possibility, Besedeš and colleagues assessed the impact of using heuristics on various types of decisions at different ages.[26] The researchers grouped participants into three age categories for the analysis: 18 to 40, 41 to 60, and older than 60.

The experiment design started with a deck of 100 cards. Each card revealed a number from one to six. Participants were given the distribution of each number in the 100-card deck. For example, they knew that the deck contained 24 of the #1 cards, 8 #2 cards, 21 #3 cards, 26 #4 cards, 12 #5 cards, and 9 #6 cards. Participants selected a set of numbers from the groups offered for the game. When a card was randomly picked, and they had that number in their group, they got $1. The task was to select the best set of numbers from several options: group A had cards 1, 3, 5, and 6; group B included 2, 3, 5, and 6; group C consisted of 1, 3, and 4; and group D held 1, 4, and 5. Groups A and B provided more numbers than group C and D, four versus three. So, if a participant used a heuristic that simply counted the numbers, this person would pick group A or B. However, the distribution information of the numbers was available. Using the distribution information showed that by selecting group A the chances of winning were 66 out of 100. Group B had 50 chances to win; group C had 71 chances; and group D had 62 chances. Thus, the optimal choice is group C. In this example, people chose between four groups (A, B, C, and D). Other variations of the experiment included 13 groups.

Not surprisingly, every age group had more difficulty finding the optimal option when provided with more groups from which to choose. Picking a needle out of a larger haystack is more difficult than from a smaller one. Figure 13.4 shows that decision making deteriorated with age. For example, 52 percent of the participants under the age of 40 selected the optimal option, but only 32 percent of those over the age of 60 did so. In some scenarios, there was a second-best group that was nearly as good as the optimal choice. The figure shows the category of "nearly optimal" for people that selected either the optimal or a nearly optimal group. The proportion of people selecting nearly optimal groups also declined with age, i.e., 72 percent to 59 percent. Older subjects had a tendency to discard some critical information about the relative importance of each option. Instead, they tended to select groups with the largest amount of numbers, rather than the groups with the highest chances to win. This approach is akin to picking a mutual fund company because it has the most funds from which to choose and ignoring the fund costs and performance. The inability of the older group to select the optimal group of numbers resulted from using heuristics. Older individuals were more easily manipulated through presentation framing and the design of the groups. Adding more groups also disproportionately affected the elderly. Their conclusions were robust to participant education level and wealth.

In summary, using heuristics is related to poor decision making that occurs as people get older. Seniors examine less information, consider fewer

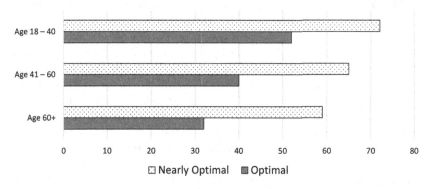

Figure 13.4 Percent Selecting Optimal or Nearly Optimal Group by Age

options when making choices, and tend to count the number of positive attributes provided by each option.

Improving Decisions of Older Adults

Since decision architecture can influence peoples' responses, can decisions be framed such that older adults perform better? The research suggests that age differences in risky decision making may be related to learning limitations of seniors operating in a new environment. Can targeted and familiar decision aids improve their financial risk-taking decisions? Samanez-Larkin and colleagues found that simple presentation of all trial expected values improved decision making in both younger and older adults.[27] Providing expected value information improved the performance of older adults to the level of matching that of pre-decision aid younger adult decisions. On the other hand, Westbrook and colleagues reported evidence that behavioral interventions may have limited effectiveness in seniors. In their experiment, providing expected value information in repeated trials evoked less behavioral change in older people than in younger adults.[28] They attributed this to the tendency for older adults to shift away from the suggested strategy over time. Although specific and reliable decision aid strategies have not been developed, research suggests that simpler and satisficing strategies are most likely to help.[29]

Use It or Lose It

Can cognitive aging be delayed? Can a cognitively demanding environment inhibit cognitive decline? Retirees are often encouraged to maintain an engaged lifestyle with intellectually stimulating activities. How challenging does an environment need to be in order for you to maintain cognitive function? Indeed, many games and puzzles claim to keep you cognitively challenged and reduce cognitive decline. Is daily use of these games enough or do you need to be fully involved in a work environment?

To address the work environment angle, one interesting study by Rohwedder and Willis examined the role of retirement in cognitive decline.[30] Through government programs and cultural differences, people tend to retire at different ages in different countries. For example, comparing the employment rate of men aged 60 to 64 years with the employment rate of men aged 50 to 54 years shows that in the United States men in their early sixties are employed at a rate that is one-third lower than men in their early fifties. In other words, about one-third of American men retire by their early sixties. However, the employment rate is over 80 percent lower for men in their early sixties in France and Austria than for men in their early fifties. That is, most men from these countries had retired by their early sixties. By examining the employment rate and cognitive surveys in the United States, England, and 11 European countries, the authors concluded that early retirement was associated with accelerated cognitive decline. Thus, remaining in a cognitively stimulating work environment reduces the impact of cognitive aging.

Can you retire and play stimulating games to keep your cognitive function high? There is much literature on this question. Simons and colleagues reviewed this literature.[31] They concluded that there is evidence that playing a cognitively stimulating game resulted in improved performance in the playing of that game. They found only some evidence of improvement in closely related tasks. Unfortunately, they reported little evidence that brain training improved everyday cognitive performance.

Summary

Populations throughout the world are aging. In the United States, the baby boom generation represents nearly 74 million people who are retired or retiring. With the decline in employer-sponsored defined benefit pensions, these retirees are being asked to manage their retirement income from personal investments. How suited are they to be successful? Due to cognitive aging, they face increasing limitations on their financial decision-making abilities. The decline in cognitive ability begins in early adulthood and continues throughout life. However, financial experience increases rapidly through the middle years. Thus, young adults are prone to making mistakes because of a lack of experience, while seniors are prone to mistakes because of cognitive decline. Estimates of peak financial decision making range from the forties to the early fifties. This is known as the U-shape of financial mistakes over time.

The cognitive decline for seniors impacts their ability to manage investments in several ways. It is shown that seniors lose some of their financial literacy, which negates their experience. While seniors appear to be aware of their cognitive decline, they remain confident about their ability to manage their finances. Cognitive ability does not appear to be related to the likelihood of obtaining financial advice, but when seniors do seek help, higher ability seniors seek better quality financial advice. Cognitive aging negatively impacts seniors' credit choices and credit ratings. Cognitive decline makes seniors more susceptible to financial fraud. Seniors become more financially risk averse over and above changes in investing needs. Declining cognitive ability results in poor stock picking, market timing, and diversification, resulting in portfolio underperformance. Finally, cognitive decline impairs a retiree's ability to manage emotions and behavioral biases in the investment decision process and causes them to make poor decisions after stock market shocks.

Questions

1   What is cognitive aging?
2   How do financial knowledge and cognitive ability interact over time?
3   How is cognitive aging associated with financial fraud?
4   How is cognitive aging related to financial decisions?

Notes

1 James M. Poterba. 2014. Retirement security in an aging population. *American Economic Review: Papers & Proceedings* 104:5, 1–30.

2 Keith Jacks Gamble, Patricia A. Boyle, Lei Yu, and David A. Bennett. 2015. Aging and financial decision making. *Management Science* 61:11, 2603–2610.

3 Sumit Agarwal, John C. Driscoll, Xavier Gabaix, and David Laibson. 2009. The age of reason: Financial decisions over the life cycle with implications for regulation. *Brookings Papers on Economic Activity* 40:2, 51–117.

4 Brian Knutson, Charles M. Adams, Grace W. Fong, and Daniel Hommer. 2001. Anticipation of increasing monetary reward selectively recruits nucleus accumbens. *Journal of Neuroscience* 21:RC159, 1–5; Gregory R. Samanez-Larkin, Sasha E. B. Gibbs, Kabir Khanna, Lisbeth Nielsen, Laura L. Carstensen, and Brian Knutson. 2007. Anticipation of monetary gain but not loss in healthy older adults. *Nature Neuroscience* 10:6, 787–791.

5 Gregory R. Samanez-Larkin, Camelia M. Kuhnen, Daniel J. Yoo, Brian Knutson. 2010. Variability in nucleus accumbens activity mediates age-related suboptimal financial risk taking. *Journal of Neuroscience* 30:4, 1426–1434.

6 Lisbeth Nielsen, Brian Knutson, and Laura L. Carstensen. 2008. Affect dynamics, affective forecasting, and aging. *Emotion* 8:3, 318–330.

7 Laura L. Carstensen, Bulent Turan, Susanne Scheibe, Nilam Ram, Hal Ersner-Hershfield, Gregory R. Samanez-Larkin, Kathryn Brooks, and John R. Nesselroade. 2011. Emotional experience improves with age: Evidence based on over 10 years of experience sampling. *Psychology and Aging* 26:1, 21–33.

8 Gregory R. Samanez-Larkin and Brian Knutson. 2014. Reward processing and risky decision making in the aging brain. In V. Reyna and V. Zayas (Eds.), Bronfenbrenner Series on the Ecology of Human Development. *The Neuroscience of Risky Decision Making*. Washington, DC, American Psychological Association.

9 Gregory R. Samanez-Larkin, Rui Mata, Peter T. Radu, Ian C. Ballard, Laura L. Carstensen, and Samuel McClure. 2011. Age differences in striatal delay sensitivity during intertemporal choice in healthy adults. *Frontiers in Neuroscience* NOV:126.

10 S.M. Hadi Hosseini, Maryam Rostami, Yukihito Yomogida, Makoto Takahashi, Takashi Tsukiura, and Ryuta Kawashima. 2010. Aging and decision making under uncertainty: Behavioral and neural evidence for the preservation of decision making in the absence of learning in old age. *NeuroImage* 52:4, 1514–1520.

11 Gregory R. Samanez-Larkin, Sara M. Levens, Lee M. Perry, Robert F. Dougherty, and Brian Knutson. 2012. Frontostriatal white matter integrity mediates adult age differences in probabilistic reward learning. *Journal of Neuroscience* 32:15, 5333–5337.

12 Michael S. Finke, John S. Howe, and Sandra J. Huston. 2017. Old age and the decline in financial literacy. *Management Science* 63:1, 213–230.

13 Annamaria Lusardi, Olivia S. Mitchell, and Vilsa Curto. 2014. Financial literacy and financial sophistication in the older population. *Journal of Pension Economics and Finance* 13:4, 347–366.

14 Keith Gamble, Patricia Boyle, Lei Yu, and David Bennett. 2015. Aging and financial decision making. *Management Science* 61:11, 2603–2610.

15 Hugh H. Kim, Raimond Maurer, and Olivia S. Mitchell. 2019. How cognitive ability and financial literacy shape the demand for financial advice at older ages. NBER Working Paper No. 2575.

16 Sumit Agarwal, John C. Driscoll, Xavier Gabaix, and David Laibson. 2009. The age of reason: Financial decisions over the life-cycle with implications for regulation. *Brookings Papers on Economic Activity* Fall, 51–117.

17 Sumit Agarwal and Bhashkar Mazumder. 2013. Cognitive abilities and household financial decision making. *American Economic Journal: Applied Economics* 5:1, 193–207.

18 Ye Li, Jie Gao, A. Zeynep Enkavi, Lisa Zaval, Elke U. Weber, and Eric J. Johnson. 2015. Sound credit scores and financial decisions despite cognitive aging. *PNAS* 112:1, 65–69.

19 APCO Insight. 2012. Senior financial exploitation study. Certified Financial Planner Board of Standards, November. Available at www.cfp.net/docs/news-events—supporting-documents/senior-americans-financial-exploitation-survey.pdf?sfvrsn=0.

20 Patricia A. Boyle, Lei Yu, Robert S. Wilson, Keith Gamble, Aron S. Buchman, and David Bennett. 2012. Poor decision making is a consequence of cognitive decline among older persons without Alzheimer's disease or mild cognitive impairment. *PLoS ONE* 7:8, 1–5.

21 Keith Jacks Gamble. 2017. Challenges for financial decision making at older ages. In Olivia S. Mitchell, P. Brett Hammond, and Stephen P. Utkus (Eds.), *Financial Decision Making and Retirement Security in an Aging World*, 33–45, Oxford, Oxford University Press.

22 Eun Jin Kim, Sherman D. Hanna, Swarn Chatterjee, and Suzanne Lindamood. 2012. Who among the elderly owns stocks? The role of cognitive ability and bequest motive. *Journal of Family and Economic Issues* 33:3, 338–352.

23 Eric Bonsang and Thomas Dohmen. 2015. Risk attitude and cognitive aging. *Journal of Economic Behavior & Organization* 112:C, 112–126.

24 George M. Korniotis and Alok Kumar. 2011. Do older investors make better investment decisions? *Review of Economics and Statistics* 93:1, 244–265.

25 Chris Browning and Michael Finke. 2015. Cognitive ability and the stock reallocations of retirees during the great recession. *The Journal of Consumer Affairs* 49:2, 356–375.

26 Tibor Besedeš, Cary Deck, Sudipta Sarangi, and Mikhael Shor. 2012. Age effects and heuristics in decision making. *Review of Economics and Statistics* 94:2, 580–595.

27 Gregory R. Samanez-Larkin, Anthony D. Wagner, and Brian Knutson. 2011. Expected value information improves financial risk taking across the adult life span. *Social Cognitive and Affective Neuroscience* 6:2, 207–217.

28 Adam Westbrook, Bruna S. Martins, Tal Yarkoni, and Todd S. Braver, T. S. 2012. Strategic insight and age-related goal-neglect influence risky decision-making. *Frontiers in Neuroscience* 6:68, 1–13.

29 Darrell A. Worthy and W. Todd Maddox. 2012. Age-based differences in strategy use in choice tasks. Frontiers in Neuroscience 5:145, 1–10; Rui Mata and Ludmila Nunes. 2010. When less is enough: Cognitive aging, information search, and decision quality in consumer choice. *Psychology and Aging* 25:2, 289–298.

30 Susann Rohwedder and Robert J. Willis. 2010. Mental retirement. *Journal of Economic Perspectives* 24:1, 119–138.

31 Daniel J. Simons, Walter R. Boot, Neil Charness, Susan E. Gathercole, Christopher F. Chabris, David Z. Hambrick, and Elizabeth A. L. Stine-Morrow. 2016. Do "brain-training" programs work? *Psychological Science in the Public Interest* 17:3, 103–186.

Index

absenteeism 43, 111, 113
Adams, R.B., Barber, B.M., and Odean, T. 44, 57n8
Add Health project 25
Addoum, J.M., Korniotis, G., and Kumar, A. 110, 118n21
Adolphs, R., Tranel, D., Damasio, H., and Damasio, A. 77n11
adoption: adoptees, investments by 4; human genome, investments of adoptees and 28–34
adoption studies, introduction to 29–32
adoptive regression 29
adrenocorticotropic hormone (ACTH) 80
affect and ethical decision making 132
affective cues and endowment effect 125
affective states and risk taking in gain/loss decision frames 128
Agarwal, S. and Mazumder, B. 199n17
Agarwal, S., Driscoll, J.C., Gabaix, X., and Laibson, D. 198n3, 198n16
Agarwal, V., Ghosh, P., and Zhao, H. 148, 151n46
age, stock market participation and 20
aggregate mood, stock market and 132–4
aging populations 183
Agnew, J., Balduzzi, P., and Sunden, A. 46, 57n16
Agrawal, A. and Lim, Y. 111–12, 119n25
agreeableness 152–3
Ahern, K.R. 147, 151n44
Ahmed, S., Sihvonen, J., and Vähämaa., S. 91, 94n25
air pollution: cognition, and financial decisions 142–5; cognition and 142–3; cognitive function and 8; on date of purchase, underperformance of stocks based on 144–5; stock market and 143–5

air quality, daily stock return and 143
Aktas, N., De Bodt, E., Bollaert, H., and Roll, R. 162, 168n43
allergies and financial decisions 145
Almenberg, J. and Dreber, A. 53, 58n32
alpha by testosterone exposure 90
ambiguity, brain function and 65–6
Amsterdam Exchange Index 140
amygdala 62–3
analytical and intuitive tendencies, measurement of 174–5
analytical versus intuitive thinking, investing decisions and 173–4
AncestryDNA 28
Andrade, E.B., Odean, T., and Lin, S. 129, 135n10
Angrisani, M., Kapteyn, A., and Lusardi, A. 52–3, 58n31
Antoine J. Bruguier, Steven R. Quarz, and Peter Bossaerts. 178–9, 182n15
Antoniou, C., Harris, R.D.F., and Zhang, R. 66, 77n7
APCO Insight 199n19
Apicella, C.L., Dreber, A., and Campbell, B. 81, 92n3
Arent, S.M., Landers, D.M., and Etnier, J.L. 118n15
asset holdings, health and 109
asset pricing bubbles, testosterone and 88
automatic (experiential) decision processes 65

Babiak, P. and Hare, R.D. 158, 167n28
Babiak, P., Neumann, C.S., and Hare, R.D. 167n37
Balloon Analog Risk Task (BART): brain function, financial decisions and 65–6; hormones, influence on

financial risk taking 83; sleep, coffee and investing 98, 102
Bannier, C.E. and Schwarz, M. 53, 58n34
Barber, B.M. and Odean, T. 48, 49, 57n20
Barel, E., Shahrabani, S., and Tzischinsky, O. 93n12
Barnea, A., Cronqvist, H., and Siegel, S. 18, 26n7
Barth, D., Papageorge, N.W., and Thom, K. 41n16
Barua, A., Davidson, L.F., Rama, D.V., and Thiruvadi, S. 49, 57n25
Bassi, A., Colacito, R., and Fulghieri, P. 138, 149n7
Bayer, C. 158
Bear, S., Rahman, N., and Post, C. 49
Beauchamp, J.P., Cesarini, D., Johannesson, M. et al. 41n13
Bechara, A., Damasio, H., Damasio, A. R., and Lee, G.P. 68, 77n12
Becker, J., Medjedovic, J., and Merkle, C. 158, 167n22
Beckmann, D. and Menkhoff, L. 47, 57n19
behavior: competitive behavior 51–2; deceptive behavior, psychopathy and 161–2; natural disasters and 145–6; terrorism and 147
behavioral biases 22–4; aging and 194–6; cognitive reflection and 175–8; investment and avoidance of 37
behavioral finance: biological foundations of 69–73; biology and psychology in finance 1–2
Behavioral Risk Factor Surveillance System (BRFSS) surveys 111, 115
Bellesi, M., De Vivo, L., Chini, M., Gilli, F., Tononi, G., and Cirelli, C. 95, 105n5
Bem's Sex Role Inventory 51
Benjamin, D.J., Cesarini, D., Van der Loos, M.J.H.M. et al. 38, 41n17–18
Bennett, B. and Wang, Z. 147, 151n43
Bernasek, A. and Shwiff, S. 46, 57n14
Bernile, G., Bhagwat, V., and Rau, P.R. 146, 151n41
Besedeš, T., Deck, C., Sarangi, S., and Shor, M. 194–5, 199n26
The 100 Best Companies to Work For in America (Levering, R., Moscowitz, M., and Katz, M.) 112
bid-ask spread, SAD and 140–41

Bierut, Laura Jean 41n12
big data 10
big five personality traits: biology and psychology in finance 8; financial decisions and 152–8; personal finances and 158–9; personality and investing 152–3
biology and psychology in finance 1–10; adoptees, investments of 4; air pollution, cognitive function and 8; behavioral finance 1–2; big data 10; big five personality traits 8; biology, behavior and 2, 10; brain function, financial decisions and 5–6; cognitive aging, diminished decision making and 9–10; cognitive biases 1–2; cognitive outcomes 7; cognitive reflection 7; cortisol 8; dark triad personality traits 9; DNA (deoxyribonucleic acid) 2–3, 4, 10; emotional and moody investors 7–8; endowment effect 2; environmental factors in financial decision making 8; exercise, diet and 6–7; extroversion 8; financial theory, evolution of 1; fraternal twins 3; functional magnetic resonance imaging (fMRI) 5; future possibilities 10; gender differences, investments and 4–5; genetic code 2; health status, decision making and 2; hormones, influence on financial risk taking 6; human genome, investments of adoptees and 4; identical twins 3; IKEA effect 2; intelligence and investment performance 9; intelligence quotient (IQ) 7, 9; mental health 7; moody investors 7–8; narcissism 9; nature *versus* nurture 2–3; neuroticism 8; norepinephrine 8; nurture environment 3; optimism 9; overconfidence 9; personality and investing 8–9; physiology 5; research gaps 10; risk aversion 8; self-interest 1; sleep, coffee and investing 6; sleep deprivation 6; technology 10; testosterone 6; traditional theories of finance 7; twin investment behavior 3; wellness, influence on financial decisions 6–7
Björklund, A., Jäntti, M., and Solon, G. 31–2, 41n8
Björklund, A., Lindahl, M., and Plug, E. 30, 41n7
Black, S.E., Devereux, P.J., Lundborg, P., and Majlesi, K. 32, 41n9

Blanchard, A. and Lyons, M. 160
Blau, B.M., DeLisle, J.R., and Price, S. M. 133–4, 136n18
Bogan, V.L. and Fertig, A.R. 113–14, 119n30–31
bond fund allocation, rebalancing frequency and 130
Bonnet M.H. and Arand, D.L. 96, 105n11
Bonow, R.O. and Eckel, R.H. 117n10
Bonsang, E. and Dohmen, T. 192, 199n23
Booth, A. and Nolen, P. 55, 58n39
Bosch-Domènech, A., Brañas-Garza, P., and Espín, A.M. 92, 94n29
Bose, S., Ladley, D., and Xin 88, 93n19
Boyle, P.A., Yu, L., Wilson, R.S. et al 190, 199n20
brain function, financial decisions and 61–78; ambiguity 65–6; amygdala 62–3; automatic (experiential) decision processes 65; Balloon Analog Risk Task (BART) 65–6; behavioral finance, biological foundations of 69–73; biology and psychology in finance 5–6; brain 61–5; brain anatomy 61–4; brain damage, evidence from 68–9; cerebellum 61–2; cingulate gyrus 62–3; controlled decision processes 65; decision processes, types of 64–5; dopamine 64; dopamine receptor D4 (DRD4) 73–4; DRD4, financial decision making and 74–5; economic decisions, brain and 65–73; electroencephalogram (EEG) 65; endowment effect 71–2; "fictive learning" 73; financial risk and reward 65–7; fornix 62–3; framing effect 70–71; frontal lobe 61–2; functional magnetic resonance imaging (fMRI) 65, 66, 67, 69, 72–3; functions and locations in brain 63; genes 73–4; genoeconomics 73–6; genoeconomics research 73; hippocampus 62–3; home ownership, financial decisions and 67; Iowa Gambling Task 68; irrational investments 72; limbic system 62–3; loss aversion 68–9; loss aversion, emotions and 70; MAOA, financial decision making and 75–6; market bubbles 66–7; medulla oblongata 61–2; monoamine oxidase (MAOA) 74, 75–6; neural activity, examination of 65; neural activity, portfolio choice and 71; neural responses to gains and losses 69–70; neuroeconomics 61; neurotransmitters 64; nucleus accumbens 63–4; occipital lobe 61–2; parietal lobe 61–2; pons 61–2; positron emission tomography (PET) 65; psychological bias 72; questions 76; receptors 73–4; serotonin 64; serotonin transporter 5-HTTLPR 74; stock selection 66; striatum 62; summary 76; temporal lobe 61–2; thalamus 62–3
Brau, B., Brau, J., Holmes, A., and Ringold, M. 163, 168n47
Breiter, H.C., Aharon, I., Kahneman, D., Dale, A. and Shizgal, P. 69–70, 77n16
Brennan, J. 112
Bressan, S., Pace, N., and Pelizzon, L. 110, 118n20
Brooks, P. and Zank, H. 45, 57n12
Brown, S. and Taylor, K. 158, 167m23
Browning, C. and Finke, M. 194, 199n25
Bucciol, A. and Zarri, L. 157, 167n18
Buffet, W. 169, 181n1
business: dark triad and 159–60; non-cognitive abilities and 163
Byrnes, J.P., Miller, D.C., and Schafer, W.D. 56n1

Cadena, B.C. and Keys, B. 165, 168n54
Cai, J., Fan, M., Ko, C.Y., Richione, M., and Russo, N. 100, 106n23
Cairney, J., Corna, L., Veldhuizen, S., Kurdyak, P., and Streiner, D.L. 108, 117n7
Calvent, L.E. and Sodini, P. 20, 27n8
Campell, K. and Minguiez-Vera, A. 49, 57n23
Capital Asset Pricing Model 140
Cappuccio, F.P., Elia, L.D., Strazzullo, P., and Miller, M.A. 95, 105n4
Carmon, Z. and Ariely, D. 71–2, 78n22
Carpenter, J.P., Garcia, J.R. and Lum, J.K. 74, 78n26
Carstensen, L.L., Turan, B., Scheibe, S. et al. 198n7
Carter, D.A., Simkins, B.J., and Simpson., W.G. 49, 57n22
Carvalho Júnior, C.V.de O., Cornacchione, E., Freitas da Rocha, A., and Rocha, F.T. 65, 76–7n3
Centers for Disease Control and Prevention (CDC, US) 107, 111, 112–13

cerebellum 61–2

Certified Financial Planner Board of Standards 189

Cesarini, D., Dawes, C.T., Johannesson, M., Lichtenstein, P., and Wallace. B. 17–18, 24–5, 26n6, 27n13

Cesarini, D., Johannesson, M., and Oskarsson., S. 40n2

Cesarini, D., Johannesson, M., Lichtenstein, P., Sandewall, Ö., and Wallace, B. 20, 27n9

Cesarini, D., Johannesson, M., Magnusson, P.K.E., and Wallace, B. 22, 27n11

Charness, G. and Gneezy, U. 45, 57n13

Chay, K.Y. and Greenstone, M. 150n30

Chen, Y. 143, 150n33

Chinese people, impact of air pollution index on cognitive performance in 143

Christelis, D., Jappelli, T., and Padula, M. 172–3, 181n6

cingulate gyrus 62–3

circulation testosterone, cortisol levels and 81–8

Clark, R.A., McConnell, J.J., and Singh, M. 150n20

Coates, J.M. and Herbert, J.K. 83–4, 93n14

Coates, J.M., Gurnell, M., and Rustichini, A. 89, 94n22

cognition, air pollution and 142–3

cognitive aging, diminished decision making and 183–99; aging populations 183; behavioral biases, aging and 194–6; biology and psychology in finance 9–10; Certified Financial Planner Board of Standards 189; cognitive ability, aging and 183–5; cognitive aging, financial fraud and 190–91; cognitive decline, cognitive stimulation and challenge to 196–7; confidence, financial knowledge and 186; Consumer Finance Monthly (CFM) survey 185–6; credit card balance transfer offers, use of 188–9; credit choices 188–9; decision processing, aging and 184–5; financial advice seeking, aging and 187–8; financial cognitive ability, age and 184; financial fraud, aging and 189–91; financial knowledge, aging and 185–7; financial literacy, aging and 185–9; financial risk aversion, aging and 185, 191–2; fraud susceptibility, cognitive aging

and 190–91; functional magnetic resonance imaging (fMRI) 184, 185; Health and Retirement Study (HRS) survey 186, 187, 191–2, 194; heuristics, use of 194–5; improving decisions of older adults 196; investment shock, reaction to 194; Ponzi schemes 189; portfolio construction and performance, aging and 193–4; portfolio performance, annualized abnormal return by age group 193–4; questions 197; retirement, role in cognitive decline 196–7; reward circuitry, structural integrity of 185; Rush Memory and Aging Project 187, 190; seniors' investment decisions 191–7; summary 197; Survey of Health, Ageing, and Retirement in Europe (SHARE) 192; Telephone Interview for Cognitive Status 192; time horizons, income needs and 191

cognitive biases 1–2

cognitive processing: investing and 170–80; types of 169–70

cognitive reflection: biology and psychology in finance 7; intelligence and investment performance 170, 174–5, 175–8

Cognitive Reflection Test (CRT): hormones, influence on financial risk taking 91–2; intelligence and investment performance 174, 175, 176, 177, 178, 179, 180

Cohen, A.J., Anderson, H.R., Ostra, B. et al. 150n31

Cohen, R.M., Gross, M., Nordahl, T.E. et al. 139, 149n13

competitive behavior 51–2

confidence, financial knowledge and 186

conscientiousness 152–3

Consumer Finance Monthly (CFM) survey 185–6

Consumer Finances, Surveys of (1989 and 2014) 46

Conte, A., Vittoria Levati, M., and Nardi, C. 125–6, 135n6

Conti, G. and Heckman, J.J. 108, 117n6

controlled decision processes 65

Corgnet, B., DeSantis, M., and Porter, D. 179–80, 182n16

Coricelli, G., Joffily, M., Montmarquette, C., and Villeval, M.C. 132, 136n14

corporate finance: big five personality traits and 157–8; gender in 49–50
corticotrophin releasing hormone (CRH) 79–80
cortisol: biology and psychology in finance 8; economic risk taking and 82; financial decisions and 83–8; hormones, influence on financial risk taking 80; traders' profits and losses and 84
Costa, L.G., Cole, T.B., Coburn, J. et al. 150n28
Costa P.T. and McCrae, R.R. 152, 166n2
Cotti, C. and Simon, D. 115, 119n40
Cotti, C., Dunn, R.A., and Tefft, N. 115, 119n38
credit card balance transfer offers, use of 188–9
credit choices, cognitive aging and 188–9
criminal convictions, genetics and 29
Cronqvist, H. and Siegel, S. 15, 23, 26n3, 27n12
Cronqvist, H. and Yu, F. 54, 58n37
Cronqvist, H., Münkel, F., and Siegel, S. 15, 21–2, 26n4
Cronqvist, H., Previtero, A., Siegel, S., and White, R.E. 90, 94n24
Cronqvist, H., Siegel, S., and Yu, F. 21–2, 27n10
Cross, C.P., Cyrenne, D.M., and Brown, G.R. 58n27
Cueva, C., Roberts, R.E., Spencer, T., Rani, N. et al. 86–7, 93n17

Da, Z., Engelberg, J., and Gao, P. 133, 136n16
D'Acunto, F., Hoang, D., Paloviita, M., and Weber, M. 172, 181n5
dark triad: conditions underlying 160; financial decisions and 159–63; personality and investing 159; personality traits 9
De Martino, B., Camerer, C.F., and Adolphs, R. 68, 77n13
De Martino, B., Kumaran, D., Seymour, B., and Dolan, R.J. 70, 77n17
De Neve, J.-E. and Fowler, J.H. 75–6, 78n31
De Neve, J.-E., Christakis, N.A., Fowler, J.H., and Frey, B.S. 25, 27n15
Deaton, A. 115, 119n34
deceptive behavior, psychopathy and 161–2

decision processes: aging and 184–5; types of 64–5
DeCovny, S. 167n27
Depue, R.A. and Collins, P.F. 166n5
Dessaint, O. and Matray, A. 146–7, 151n42
DeYoung, C.G., Hirsh, J.B., Shane, M. S. et al. 166n4
Dickinson, D.L., Chaudhuri, A., and Greenaway-McGrevy, R. 100–101, 106n24
Diekelmann, S. and Born, J. 105n13
disposition effect 23–4; air pollution and 145; impact of emotions on 127
DNA (deoxyribonucleic acid): biology and psychology in finance 2–3, 4, 10; human genome, investments of adoptees and 28, 34, 35, 38, 40
Dohmen, T., Falk, A., Huffman, D., and Sunde, U. 173, 181n7
Dolvin, S.D. and Pyles, M.K. 141–2, 150n27
Dolvin, S.D., Pyles, M.K., and Wu, Q. 141, 150n25
dopamine 64
dopamine receptor D4 (DRD4) 73–4
Dow Jones Industrial Average stock index, health impacts on 115
DRD4, financial decision making and 74–5
Dreber, A., Apicella, C.L., Eisenberg, D. T.A. et al. 74–5, 78n27
Dreher, J.-C., Schmidt, P.J., Kohn, P. et al. 43, 56n2
Drummond, S.P.A., Brown, G.B., Stricker, J.L. et al. 105n6
dual hormone hypothesis 82–3
Durand, R.B., Fung, L., and Limkriangkrai, M. 155, 166n12
Durand, R.B., Newby, R., and Sanghani, J. 155–6, 166n13
Durand, R.B., Newby, R., Peggs, L., and Siekierka, M. 156, 166n14
Durand, R.B., Newby, R., Tant, K., and Trepongkaruna, S. 156, 166n15
Dynamic Experiments for Estimating Preferences survey 110–11

Eagles, J.M. 149n5
econometric techniques 19–20, 21, 22
economic decisions, brain and 65–73
economic policy uncertainty, impact on healthy choices 116
Edmans, A. 112, 119n26–7

education: attainment in, genetics and 29; income and 30–31, 36
Edwards, R.D. 109, 118n17
Ekman, P. 123, 135n1
electroencephalogram (EEG) 65
electronic trading funds (ETFs) 84–5
emotional and moody investors 123–36; affect and ethical decision making 132; affective cues and endowment effect 125; affective states and risk taking in gain/loss decision frames 128; aggregate mood, stock market and 132–4; bet average amount based on treatment and experience 131; biology and psychology in finance 7–8; bond fund allocation, rebalancing frequency and 130; disposition effect, impact of emotions on 127; emotion, origins of 123–4; emotional affective states 124–5; emotional regulation 131–2; emotions, mood, and physiology 123–4; emotions and economic decisions 124–34; endowment effect, emotions on 125; xperimental finance, emotions, mood and 124–9; investment risk, impact of emotions on 128; irrational investment decisions 123; loss aversion and affect 129–32; mood positivity, investment decisions and 133–4; myopic loss aversion 129–30; physiological response to economic decisions 124; prospect theory predictions 127; questions 135; risk preferences, effect of emotions on 125–6; risk taking, impact of fear and anger on 126–7; stock market bubbles, impact of emotion on 129; summary 134–5
employee health, company performance and 111–12
endowment effect: biology and psychology in finance 2; brain function, financial decisions and 71–2; emotions and 125
Engelberg, J. and Parsons, C.A. 116, 119n41
entrepreneurship, genetics and 29
environmental factors in financial decision making 137–51; air pollution, cognition, and financial decisions 142–5; air pollution and cognition 142–3; air pollution and stock market 143–5; air pollution on date of purchase, underperformance of stocks

based on 144–5; air quality, daily stock return and 143; allergies and financial decisions 145; Amsterdam Exchange Index 140; behavior, natural disasters and 145–6; behavior, terrorism and 147; bid-ask spread, SAD and 140–41; biology and psychology in finance 8; Capital Asset Pricing Model 140; Chinese people, impact of air pollution index on cognitive performance in 143; cognition, air pollution and 142–3; disposition effect, air pollution and 145; environment, human genome, investment decisions and 32–3; Environmental Protection Agency (EPA, US) 142; experimental finance, seasonal affective disorder (SAD) and 139; experimental finance, weather and 138; financial analysts, seasonal affective disorder (SAD) and 141–2; financial decisions, natural disasters and 146–7; hurricane strike, reactions to salient risk following 146–7; initial public offerings (IPOs), SAD and pricing of 141–2; Investor Behavioral Project Survey (Yale University) 138; Longitudinal Internet Studies for the Social Sciences (LISS) panel data 139–40; melatonin, effects of 137–8; mood, weather and 137–8; municipal bond market. impact of natural disasters on 147; natural disasters, financial decision making and 145–7; particulate matter, air pollution and 142; physiology, seasonal affective disorder (SAD) and 138–9; positron emission tomography (PET) scans 139; questions 148; respiratory system, air pollution and 142–3; seasonal affective disorder 138–42; S&P 500 returns, influences of fine particulate matter (PM2.5) on 144; stock market, air pollution and 143–5; stock market, seasonal affective disorder (SAD) and 139–41; stock market, terrorism and 148; stock market, weather and 138; summary 148; sunshine, serotonin and 137; terrorism, stock market and 147–8; weather, mood, and financial decisions 137–8
Environmental Protection Agency (EPA, US) 142

Erickson, K.I., Voss, M.W., Prakash, R. S. et al. 118n13

Eriksson, K. and Simpson, B. 50–51, 58n28

excess return, testosterone, cortisol and 85

exercise, diet and 6–7, 108

expected utility theory, risk aversion, personality and 154

experimental finance: emotions, mood and 124–9; seasonal affective disorder (SAD) and 139; weather and 138

experimental investments, big five personality traits and 155–6

experimental risk taking, big five personality traits and 154–5

extroversion: biology and psychology in finance 8; personality and investing 152–3

facial masculinity and 2D:4D ratio 88–9

Fagereng, A., Mogstad, M., and Rønning, M. 33, 41n10

female chief financial officers (CFOs) 49–50

feminism 50–52

'fictive learning' 73

Filbeck, G., Hatfield, P., and Horvath, P. 154, 166n10

finance, impact on health of 114–16

financial advice seeking, aging and 187–8

financial analysts, seasonal affective disorder (SAD) and 141–2

financial cognitive ability, age and 184

financial crisis (2008–2009), well-being during 115

financial decisions: dark triad and 160–63; natural disasters and 146–7; non-cognitive abilities and 165

financial distress, effects of 115–16

financial experiment research 44–5

financial fraud, aging and 189–91

financial information, acquisition of 53–4

financial knowledge, aging and 185–7

financial literacy 52–4; aging and 185–9

financial risk and reward 65–7

financial risk aversion, aging and 185, 191–2

financial theory, evolution of 1

financial wealth, influences of nature and nurture on 33–4

Finke, M.S., Howe, J.S., and Huston, S. J. 185–6, 198n12

Finkelstein, E.A., Trogdon, J.G., Cohen, J.W., and Dietz, W. 107, 117n3

Fisher, P.J. and Yao, R. 46, 57n18

fluid intelligence 169–70, 172–3

fornix, brain function and 62–3

Fortune 112

fragmentation of sleep 96

framing effect 70–71

Francis, B., Hasan, I., Park, J.C., and Wu, Q. 50, 58n26

fraternal twins: biology and psychology in finance 3; twin investment behavior 13–14

fraud susceptibility, cognitive aging and 190–91

Frederick, S. 174, 181n9

Friesen, W.V. 123, 135n1

frontal lobe 61–2

Frydman, C. and Camerer, C. 72–3, 78n25

Frydman, C., Barberis, N., Camerer, C., Bossaerts, P., and Rangel, A. 72, 78n24

Frydman, C., Camerer, C., Bossaerts, P., and Rangel, A. 75, 78n29

Frydman, Cary 71, 78n21

functional magnetic resonance imaging (fMRI): biology and psychology in finance 5; brain function, financial decisions and 65, 66, 67, 69, 72–3; cognitive aging, diminished decision making and 184, 185; personality and investing 153; sleep, coffee and investing 98

future possibilities 10

Gamble, K., Boyle, P., Yu, L., and Bennett, D. 187, 198n14

Gamble, K.J. 199n21

Gamble, K.J., Boyle, P.A., Yu, L., and Bennett, D.A. 198n2

Garbarino, E., Slonim, R., and Sydnor, J. 89, 93n20

Garrett, I., Kamstra, M.J., and Kramer, L.A. 140, 149n18

gender differences, investments and 42–58; biology and psychology in finance 4–5; competitive behavior 51–2; Consumer Finances, Surveys of (1989 and 2014) 46; corporate finance, gender in 49–50; defined contribution retirement plans, gender differences in 46; female chief financial officers (CFOs) 49–50; feminism 50–52;

financial experiment research 44–5; financial information, acquisition of 53–4; financial literacy 52–4; gender biology 43; gender differences, determinants of 42–4; gender of Others 54–6; gender research 42; household financial decisions, impact of gender norms on 54; household obligations of women 44; investment, impact of gender on 45; *Jeopardy* (TV game show) 44; loss aversion behavior 45; masculinity 50–52; mutual fund choice, gender and 48; National Financial Capability Study 52–3; nature, social norms or? 50–54; neural differences 43; parental financial risk taking, impact of having a female baby on 54–5; questions 56; retail investment, impact of gender on 49; retirement accounts 45–6; risk aversion of fund managers 47; risk aversion scores of parents 55; risk taking, impact of gender on 44–5; risk-taking behavior, impact of the environment on 55–6; risky portfolio holdings in Finland, gender differences in 53; sensation-seeking behavior 45; sex 42, 51; social risk 42; societal differences 43–4; STEM (Science, Technology, Engineering, and Math) careers, underrepresentation of women in 43–4; stock market participation, actual and perceived financial literacy in 53; stock market participation, financial literacy and gender for 53; stock market participation, gender and 20; summary 56; trading behavior 46–9; trading decisions, impact of gender on 48–9; Wall Street, women on 44; Zuckerman's Sensation Seeking Scale 50
General Educational Development (GED) 163
genes 73–4
genetic code 2
genetic contribution to life satisfaction 25
genetic differences, savings behavior and 15
genetic variants 35
genetics: behavioral biases and 22–3; earnings and 14; investment choices and 24; investment decisions and 32–3; pension choices and 21
genoeconomics 73–6; research in 73

genome 34–9
genome-wide association studies (GWAS) 35, 38
genomic-relatedness-matrix restricted maximum likelihood (GREML) 38, 39
Genos 28
Giulietti, C., Tonin, M., and Vlassopoulos, M. 115, 119n39
giving 24–5
global market games 101
Gneezy, U. and Potters, J. 81, 92n2
Goetzmann, W.N., Kim, D., Kumar, A., and Wang, Q. 138, 149n9
Goldman, D. and Maestas, N. 110, 118n19
Gow, I.D., Kaplan, S.N., Larcker, D.F., and Zakolyukina, A.A. 157, 167n19
Graham, Benjamin 21, 22
Green, C., Jegadeesh, N., and Tang., Y. 44, 57n9
Greene, F.J., Han, L., Martin, S., Zhang, S., and Wittert, G. 82, 93n6
Gregory-Allen, R., Jacobsen, B., and Marquering, W. 100, 106n22
Grey, John 4, 10n3
Grinblatt, M. and Keloharju, M. 49, 57n21
Grinblatt, M., Ikäheimo, S., Keloharu, M., and Knüpfer, S. 171, 181n4
Grinblatt, M., Keloharju, M., and Linnainmaa, J.T. 170–71, 181n2–3
grittiness, retirement preparation, financial capability and 158–9
Guthrie, K. and Sokolowsky, J. 107, 117n5

Haigh, M.S. and List, J.A. 130, 135n12
Halko, M.-L., Kaustia, M., and Alanko, E. 53, 58n33
Ham, C., Lang, M., Seybert, N., and Wang, S. 162, 168n42
Hammond, R.A. and Levine, R. 107, 117n4
Hannak, A., Anderson, A., Feldman Barrett, L. et al. 137, 149n2
happiness 25
Hare, R.D. 167n26
Hasler, A. and Lusardi, A. 57n6
health: obesity and 107–8; risk taking and, experimental evidence 110–11; status of, decision making and 2
Health and Retirement Study (HRS): cognitive aging, diminished decision

making and 186, 187, 191–2, 194; human genome, investments of adoptees and 36–7; personality and investing 157; wellness, influence on financial decisions 109, 110, 113, 114, 115

Heaton, J.B. 164, 168n52

Heckman, J.J. and Rubinstein, Y. 163, 168n46

Heckman, J.J., Stixrud, J., and Urzua, S. 168n45

hedge fund management, psychopathic behavior and 162–3

Hennig, J. 166n6

Hermans, E.J., Ramsey, N.F., and Van Honk., J. 93n11

heuristics, use of 9, 170, 174, 175, 180, 194–5

Heyes, A., Neidell, M., and Saberian, S. 144, 150n36

high-fee funds, IQ and avoidance of 171–2

Hill, D.L. 160, 167n36

Hillman, C.H., Erickson, K.I., and Kramer. A.F. 118n13

hippocampus 62–3

Hirshleifer, D. and Shumway, T. 138, 149n8

Hirshleifer, D., Low, A., and Teoh, S.H. 164, 168n51

Hjalmarsson, R. and Lundquist, M.J. 40n4

Holt and Laury lottery task. 138, 154

Holt International Children's Services 29

homeownership: financial decisions and 67; twin investment behavior 15–16

Hoppe, E.I. and Kusterer, D.J. 175–6, 181n11

hormonal physiology 79–80

hormones, influence on financial risk taking 79–94; adrenocorticotropic hormone (ACTH) 80; alpha by testosterone exposure 90; asset pricing bubbles, , testosterone and 88; Balloon Analog Risk Task (BART) 83; biology and psychology in finance 6; circulation testosterone, cortisol levels and 81–8; Cognitive Reflection Test (CRT) 91–2; corticotrophin releasing hormone (CRH) 79–80; cortisol 80; cortisol, economic risk taking and 82; cortisol, financial decisions and 83–8; cortisol, traders' profits and losses and 84; dual hormone hypothesis 82–3; electronic trading funds (ETFs) 84–5; excess return, testosterone, cortisol and 85; facial masculinity and 2D:4D ratio 88–9; hormonal physiology 79–80; hormones 79–80; investment biases, testosterone, cortisol and 86; Iowa Gambling Task 81; questions 92; risk taking, testosterone, cortisol and 87; Rotman Interactive Trader 84–5; Russian Longitudinal Monitory Survey 89; summary 92; Swedish Twin Registry 90; testosterone 79–80; testosterone, economic risk taking and 81–2; testosterone, financial decisions and 83–8; testosterone, traders' profits and losses and 83; testosterone proxies 88–92; testosterone proxies, economic risk taking and 89; testosterone proxies, financial decision making and 89–91; testosterone proxies, validity of 91–2; 2D:4D ratio, facial masculinity and 88–9

Hosseini, S.M.H., Rostami, M., Yomogida, Y., Takahashi, M., Tsukiura, T., and Kawashima, R. 198n10

household financial decisions, impact of gender norms on 54

household obligations of women 44

Howarth, E. and Hoffman, M.S. 137, 149n1

Hrazdil, K., Novak, J., Rogo, R., Wiedman, C.I., and Zhang, R. 157, 167n20

Hsu, M., Bhatt, M., Adolphs, R., Tranel, D., and Camerer, C.F. 66, 77n6

Huang, J., Xu, N., and Yu, H. 144–5, 151n38

Huettel, S.A., Stowe, C.J., Gordon, E. M., Warner, B.T., and Platt, M.L. 77n5

human genome, investments of adoptees and 28–41; adoption 28–34; adoption studies, introduction to 29–32; adoptive regression 29; AncestryDNA 28; behavioral biases, investment and avoidance of 37; biology and psychology in finance 4; criminal convictions, genetics and 29; DNA (deoxyribonucleic acid) 28, 34, 35, 38, 40; education, income and 30–31, 36; educational attainment, genetics and 29; entrepreneurship, genetics and 29;

environment, investment decisions and 32–3; financial wealth, influences of nature and nurture on 33–4; genetic variants 35; genetics, investment decisions and 32–3; genome 34–9; genome-wide association studies (GWAS) 35, 38; genomic-relatedness-matrix restricted maximum likelihood (GREML) 38, 39; Genos 28; Health and Retirement Study (HRS) 36–7; Holt International Children's Services 29; Human Genome Project 28, 34; investing decisions 32–4; investing practice 36–8; longevity, genetics and 29; macroeconomy, risks and uncertainties associated with 37–8; nature, nurture, and unique experiences, variance decomposition attributable to 29–30; nature and nurture, importance within family types 31–2; Norway, adoptee database of Korean infants in 33; questions 40; risk attitudes, genetic contribution to 38–9; risk aversion 38–9; risk aversion, educational attainment polygenic score and 37; single nucleotide polymorphisms (SNPs) 34–5, 36, 38, 40; S&P 500 Index 37; summary 39–40; 23andMe 28; unique experiences, variance decomposition attributable to 29–30; Veritas Genetics 28; voting behavior, genetics and 29; wealth, intergenerational transmission of 33; wealth in retirement, educational attainment polygenic score and 36–7
hurricane strike, reactions to salient risk following 146–7

Ichino, A. and Moretti, E. 43, 57n3
identical (monozygotic) twins 3, 13
Iger, R. 112
IKEA effect 2
impatience, cognitive ability and 173
Income Dynamics, Panel Study of 114
inflation expectations 172
Ingalhalikar, M., Smith, A., Parker, D. et al. 57n5
initial public offerings (IPOs), SAD and pricing of 141–2
intelligence and investment performance 169–82; analytical and intuitive tendencies, measurement of 174–5;

analytical *versus* intuitive thinking, investing decisions and 173–4; behavioral biases, cognitive reflection and 175–8; biology and psychology in finance 9; cognitive processing, investing and 170–80; cognitive processing, types of 169–70; cognitive reflection 170, 174–5, 175–8; Cognitive Reflection Test (CRT) 174, 175, 176, 177, 178, 179, 180; conservatism bias 176–7; fluid intelligence 169–70, 172–3; high-fee funds, IQ and avoidance of 171–2; impatience, cognitive ability and 173; inflation expectations 172; intelligence types, impacts on investing decisions of 179–80; intuitive and analytical tendencies, measurement of 174–5; intuitive *versus* analytical thinking, investing decisions and 173–4; investing decisions, impacts of intelligence types on 179–80; investing decisions, theory of mind and 178–9; investment returns, IQ and 171; IQ, investing decisions and 170–73; mind, theory of 170; mutual fund choices 171–2; prospect theory 177, 181; questions 181; risk aversion, cognitive ability and 173; stock market participation rate, IQ and 170–71; summary 180–81; Survey of Health, Ageing, and Retirement in Europe (SHARE) 172–3; theory of mind 170
intelligence quotient (IQ) 7, 9, 169–73, 179–80
investing decisions: human genome, investments of adoptees and 32–4; impacts of intelligence types on 179–80; theory of mind and 178–9; twin investment behavior 18–22
investing practice 36–8
investing style 21–2; preferences in, drivers of 22
investment, impact of gender on 45
investment biases, testosterone, cortisol and 86
investment returns, IQ and 171
investment risk, impact of emotions on 128
investment shock, reaction to 194
investments, big five personality traits and 156–7
Investor Behavioral Project Survey (Yale University) 138

Iowa Gambling Task: brain function, financial decisions and 68; hormones, influence on financial risk taking 81; sleep, coffee and investing 102–3, 104

IQ, investing decisions and 170–73

irrational investment decisions 72, 123

Jadlow, J.W. and Mowen, J.C. 165, 168n55

James III, R.N., and O'Boyle, M.W. 77n10

Jeopardy (TV game show) 44

Jetter, M. and Walker, J.K. 44, 57n10

Jia, Y., Van Lent, L., and Zeng, Y. 91, 94n27

Jianakoplos, N.A. and Bernasek, A. 46, 57n15

"Jim twins" 13

Johansson, C., Smedh, C., Partonen, T. et al. 149n10

Kahn, H. and Cooper, C.L. 116, 119n44

Kahn, H., Cooper, C.L., and Elsey, S.P. 116, 119n45

Kahneman, D. 173–4, 181n8

Kahneman, Daniel 1

Kalcheva, I., McLemore, P., and Sias, R. 116, 119n43

Kamiya, S., Kim, Y.H., and Suh, J. 91, 94n26

Kamstra, M.J., Kramer, L.A., and Levi, M.D. 99–100, 106n20–21, 140, 149n17

Kamstra, M.J., Kramer, L.A., Levi, M. D., and Wermers, R. 141, 150n23

Kandasamy, N., Hardy, B., Page, L. et al. 82, 93n8

Kaplanski, G., Levy, H., Veld, C., and Veld-Merkoulova, Y. 139–40, 149n16

Katz, M. 112

Ke, Da 54, 58n26

Kennis, M., Rademaker, A.R., and Geuze, E. 166n3

Kiehl, K.A., Smith, A.M., Hare, R.D. et al. 168n38

Killgore, W.D.S., Grugle, N.L., and Balkin, T.J. 102, 106n26

Killgore, W.D.S., Grugle, N.L., Killgore, D.B. et al 101–2, 106n25

Killgore, W.D.S., Lipizzi, E.L., Kamimori, G.H., and Balkin, T.J. 103–4, 106n27

Killgore, William D. 98, 105n17

Kim, E.J., Hanna, S.D., Chatterjee, S., and Lindamood, S. 191–2, 199n22

Kim, H.H., Maurer, R., and Mitchell, O.S. 187–8, 198n15

Kim, J.-B., Wang, Z., and Zhang, L. 164, 168n50

Kleinfeld, K. 112

Knutson, B., Adams, C.M., Fong, G.W., and Hommer, D. 198n4

Knutson, B., Wimmer, G.E., Rick, S. et al. 72, 78n23

Korniotis, G.M. and Kumar, A. 193, 199n24

Kountouris, Y. and Remoundou, K. 99, 106n19

Kowalski, C.M., Vernon, P.A., and Aitken Schermer, J. 167n32

Kramer, A.F. and Erickson, K.I. 118n13

Kramer, L.A. and Weber, J.M. 139, 149n15

Kuhnen, C.M. and Chiao, J.Y. 75, 78n28

Kuhnen, C.M. and Knutson, B. 69, 77n15, 124–5, 135n4

Kuhnen, C.M. and Melzer, B.T. 165, 168n53

Lam, R.W. and Levitan, R.D. 149n11

Lee, D., Seo, H., and Jung, M.W. 64, 76n2

Legere, J. 112

Lepori, G.M. 144, 150n37

Leproult, R., Copinschi, G,, Buxton, O., and Van Cauter, E. 96, 105n12

Lerner, J.S. and Keltner, D. 126, 135n7

Lerner, J.S., Small, D.A., and Loewenstein, G. 125, 135n5

Levering, R. 112

Levine, Seymour 57n4

Levy, T. and Yagil, J. 143–4, 150n35

Lewis, Jim ("Jim twin") 13

Li, J.J., Massa, M., Zhang, H., and Zhang, J. 145, 151n39

Li, R., Smith, D.V., Clithero, J.A. et al. 71, 78n20

Li, Y., Gao, J., Enkavi, A.Z., Zaval, L., Weber, E.U., and Johnson, E.J. 189, 199n18

Limbach, P. and Sonnenburg, F. 112, 119n28

limbic system 62–3

Lindahl, M., Lundberg, E., Palme, M., and Simeonova, E. 40n3

Lindeboom, M. and Melnychuk, M. 114, 119n32

Lindquist, M.J., Sol, J., and Van Praag, M. 40n1
Lintner, J. 140
Liu, M. 157, 167n21
Lo, A.W. and Repin, D.V. 124, 135n3
Lo, K. and Wu, S.S. 141, 150n26
London Stock Exchange 83, 89, 116
longevity, genetics and 29
Longitudinal Internet Studies for the Social Sciences (LISS) panel data 139–40, 158
Loomis, C. 169, 181n1
loss aversion 68–9; affect and 129–32; emotions and 70; loss aversion behavior 45
Love, D.A. and Smith, P.A. 110, 118n22
Lu, Y. and Teo, M. 89–90, 94n23
Lusardi, A., Mitchell, O.S., and Curto, V. 186, 198n13
Lykken, D. and Tellegen, A. 27n14

Machiavellianism 9, 152, 159
Maciejovsky, B., Schwarzenberger, H., and Kirchler, E. 132, 136n15
macroeconomy, risks and uncertainties associated with 37–8
Malmendier, U. and Tate, G. 163–4, 168n48–9
MAOA, financial decision making and 75–6
market bubbles 66–7
Markowitz, Harry 16
Marshall, Alfred 1
masculinity 50–52
Mayew, W.J. and Venkatachalam 134, 136n20
McInerney, M., Mellor, J.M., and Nicholas, L.H. 115, 119n33
McKenna, B.S., Dickinson, D.L., Orff, H.J., and Drummond, S.P.A. 97–8, 105n16
McTier, B.C., Tse, Y., and Wald, J.K. 112–13, 119n29
mean-variance portfolio theory (Markowitz) 16
medulla oblongata 61–2
Mehta, P. and Josephs., R.A. 93n10
Mehta, P.H., Welker, K.M., Zilioli, S., and Carré, J.M. 83, 93n13
Meier-Pesti, K. and Penz, E. 51, 58n29
melatonin, effects of 137–8
memory, deep sleep and 96

Men Are from Mars, Women Are from Venus (Grey, J.) 4, 10n3
mental health 7; financial decisions and 113–14
Merriam-Webster.com 166n1
mind, theory of 170, 178–80
Minnesota Center for Twin and Family Research 14
Minnesota Twin Registry 16
Mitler, M.M., Carskadon, M.A., Czeisler, C.A. et al. 104n1
Molin, J., Mellerup, E., Bolwig, T., Scheike, T., and Dam, H. 139, 149n14
Molteni, R., Barnard, R., Ying, Z., Roberts, C.K., and Gomez-Pinilla, F. 117n11
monoamine oxidase (MAOA) 74, 75–6
mood: moody investors, psychology of 7–8; positivity in, investment decisions and 133–4; weather and 137–8
Morningstar portfolio management 21, 22, 53–4
Moscowitz, M. 112
Mossin, J. 140
municipal bond market. impact of natural disasters on 147
mutual fund choice: gender and 48; intelligence and investment performance 171–2
myopic loss aversion 129–30; impact of personality and gender on 155

Nadler, A., Jiao, P., Johnson, C.J., Alexander, V., and Zak, P.J. 88, 93n18
narcissism: biology and psychology in finance 9; personality and investing 159; quality of financial reporting and 162
National Financial Capability Study 52–3, 58n31
National Health and Nutrition Examination Survey (NHANES) 117n1
National Longitudinal Survey of Youth (NLSY) 165
National Postsecondary Student Aid Study (NPSAS) 165
natural disasters, financial decision making and 145–7
nature, social norms or? 50–54

nature and/or nurture: biology and psychology in finance 2–3; importance within family types 31–2; unique experiences and 13–14; unique experiences and, variance decomposition attributable to 29–30; *see also* nurture

Nave, G., Nadler, A., Zava, D., and Camerer, C. 92, 94n28

Netter, P. 166n7

neural activity: big five personality traits and 153–4; examination of, brain function and 65; portfolio choice and 71

neural differences 43

neural responses to gains and losses 69–70

neuroeconomics 61

neuroticism: biology and psychology in finance 8; personality and investing 152–3

neurotransmitters 64

Nicholson, N., Soane, E., Fenton-O'Creevy, M., and Willman, P. 154, 166n8

Niederle, M. and Vesterlund, L. 51, 58n39

Nielsen, L., Knutson, B., and Carstensen, L.L. 198n6

Nir, Y., Andrillon, T., Marmelshtein, A. et al 96, 105n9

Nofsinger, J. and Varma, A. 177, 181n12

Nofsinger, J.R. and Shank, C. 111, 118n24

Nofsinger, John R. 10n1

Nofsinger, J.R. and Shank, C.A. 96–7, 105n15

Nofsinger, J.R., Patterson, F.M., and Shank, C.A. 84–6, 93n15–16

noncognitive abilities: financial decisions and 163–5; personality, investing and 163

norepinephrine 8

Norway, adoptee database of Korean infants in 33

nucleus accumbens 63–4

nurture: environment of 3; experiences of 19; pension choices and 21; savings behavior and 15

Nye, J.V., Bryukhanov, M., Kochergina, E. et al. 89, 94n21

NYSE Internal Database and Consolidated Tape Statistics 150n19

Oberlechner, T. and Nimgade, A. 80, 92n1, 116, 119n46

obesity: health and 107–8; impact on corporate policies 111–12

occipital lobe 61–2

Oechssler, J., Roider, A., and Schmitz, P.W. 178, 181n13

Oehler, A. and Wedlich, F. 154, 166n11

Oehler, A., Wendt, S., Wedlich, F., and Horn, M. 156, 167n16

Okbay, A., Beauchamp, J.P., Fontana, M.A. et al. 41n14

Ong, Q., Theseira, W., and Ng, I.Y.H. 114–15, 119n33

openness 152–3

optimism 9; overconfidence and 164

Organization for Economic Co-operation (OECD) 107, 117n2

overconfidence 9; optimism and financial decisions 163–4

Pantzalis, C. and Ucar, E. 145, 151n40

Papageorge, N.P. and Thom, K. 41n15

parietal lobe 61–2

Parise, G. and Peijnenburg, K. 158, 167m24

particulate matter, air pollution and 142

Patterson, F. and Shank, C. 110–11, 118n23

pension choices: fund choices, analysis of 21; twin investment behavior and 20–21

Perez, P.R. 168n39

personality and investing 152–68; agreeableness 152–3; big five personality traits 152–3; big five personality traits, financial decisions and 152–8; big five personality traits, personal finances and 158–9; biology and psychology in finance 8–9; business, dark triad and 159–60; business, non-cognitive abilities and 163; conditions underlying dark triad 160; conscientiousness 152–3; corporate finance, big five personality traits and 157–8; dark triad 159; dark triad, financial decisions and 159–63; deceptive behavior, psychopathy and 161–2; expected utility theory, risk aversion, personality and 154; experimental investments, big five personality traits and 155–6; experimental risk taking, big five personality traits and 154–5; extroversion 152–3;

financial decisions, dark triad and 160–63; financial decisions, non-cognitive abilities and 165; functional magnetic resonance imaging (fMRI) 153; General Educational Development (GED) 163; grittiness, retirement preparation, financial capability and 158–9; Health and Retirement Study (HRS) 157; hedge fund management, psychopathic behavior and 162–3; investments, big five personality traits and 156–7; Longitudinal Internet Study for the Social Sciences 158; myopic loss aversion, impact of personality and gender on 155; narcissism 159; narcissism, quality of financial reporting and 162; National Longitudinal Survey of Youth (NLSY) 165; National Postsecondary Student Aid Study (NPSAS) 165; neural activity, big five personality traits and 153–4; neuroticism 152–3; non-cognitive abilities 163; non-cognitive abilities, financial decisions and 163–5; openness 152–3; optimism and overconfidence 164; over-confidence, optimism, and financial decisions 163–4; Psychopathic Personality Inventory-Revised (PPI-R) survey 161; psychopathy 159, 160; questions 166; relationship between personality traits and neuro-transmitters and hormones 153–4; risk and return expectations, relationship between personality and 154–5; stock price crash risk, overconfidence and 164

Phan, K.L., Wager, T., Taylor, S.F., and Liberzon, I. 123, 135n2

physical health, financial decisions and 109–13

physiology: biology and psychology in finance 5; physiological response to economic decisions 124; seasonal affective disorder (SAD) and 138–9

Pinegar, J. Michael 99–100, 106n21

Pittsburg Sleep Quality Index 97

Plug, E. and Vijverberg, W. 40n5

Pogrebna, G., Oswald, A.J., and Haig, D. 54, 58n38

Polidori, M.C., Praticó, D., Mangialasche, F. et al 118n12

pons, brain function and 61–2

Ponzi schemes 189

portfolio choice, health and 109–10

portfolio construction and performance, aging and 193–4

portfolio holdings, behavioral biases and 23

portfolio performance, annualized abnormal return by age group 193–4

positron emission tomography (PET) 65, 139

Poterba, J.M. 198n1

Powell, M. and Ansic, D. 44, 57n11

Price, Jr., T. Rowe 21, 22

Price, S.M., Doran, J.S., Peterson, D.R., and Bliss, B.A. 133, 136n17

Price, S.M., Seiler, M.J., and Shen, J. 134, 136n19

Prinz, S., Gründer, G., Hilgers, R.-D., Holtemöller, O., and Vernaleken, I. 154, 166n9

prospect theory 16, 70, 98; intelligence and investment performance 177, 181; predictions of, emotional and moody investors and 127

psychological biases: brain function, financial decisions and 72; twin investment behavior and 22–3, 23–4

Psychopathic Personality Inventory-Revised (PPI-R) survey 161

psychopathy 159, 160

questions: brain function, financial decisions and 76; cognitive aging, diminished decision making and 197; emotional and moody investors 135; environmental factors in financial decision making 148; gender differences, investments and 56; hormones, influence on financial risk taking 92; human genome, investments of adoptees and 40; intelligence and investment performance 181; personality and investing 166; sleep, coffee and investing 104; twin investment behavior 26; wellness, influence on financial decisions 117

Ragatz, L.L., Fremouw, W., and Baker, E. 167n29

Ramon P. DeGennaro, Mark J. Kamstra, and Lisa A. Kramer. 141, 150n22

Rao, H., Korczykowski, M., Pluta, J., Hoang, A., and Detre, J.A. 77n4

rapid eye movement (REM) sleep 96

Ratcliffe, A. and Taylor, K. 115, 119n37

receptors, brain function and 73–4
Reinemund, S. 112
research gaps; biology and psychology in finance 10
respiratory system, air pollution and 142–3
retail investment, impact of gender on 49
retirement: defined contribution retirement plans, gender differences in 46; retirement accounts 45–6; role in cognitive decline 196–7; wealth in, educational attainment polygenic score and 36–7
reward circuitry, structural integrity of 185
Richards, C.H., Gilbert, G., and Harris, J.R. 167n33
Rind, B. and Strohmetz, D. 149n4
Riquelme, H.E. and Rios, R.E. 43, 57n7
risk and return expectations, relationship between personality and 154–5
risk attitudes: genetic contribution to 38–9; twin investment behavior and 16–18
risk aversion: biology and psychology in finance 8; cognitive ability and 173; educational attainment polygenic score and 37; of fund managers 47; human genome, investments of adoptees and 38–9; parents' scores 55; twin investment behavior 18
risk investment 18
risk preferences: determination of 16–17, 17–18; effect of emotions on 125–6
risk-return models 16
risk taking: impact of fear and anger on 126–7; impact of gender on 44–5; impact of the environment on 55–6; risky portfolio holdings in Finland, gender differences in 53; testosterone, cortisol and 87
Risom, L., Møller, P., and Loft, S. 150n29
Rodahl, Kaare 95, 104n2
Rohwedder, S. and Willis, R.J. 196–7, 199n30
Roiser, J.P., De Martino, B., Tan, G.C.Y. et al. 77n17
Rosen, H.S. and Wu, S. 109, 118n16
Rosenthal, N.E., Sack, D.A., Gillin, J.C. et al. 149n12
Roth, S. 143, 150n34
Rotman Interactive Trader 84–5
Rush Memory and Aging Project 187, 190
Russian Longitudinal Monitory Survey 89

Sacerdote, Bruce 40n6
Samanez-Larkin, G.R. and Knutson, B. 198n8
Samanez-Larkin, G.R., Gibbs, S.E.B., Khanna, K. et al. 198n4
Samanez-Larkin, G.R., Kuhnen, C.M., Yoo, D.J., and Knutson, B. 198n5
Samanez-Larkin, G.R., Mata, R., Radu, P.T., Ballard, I.C., Carstensen, L.L., and McClure, S. 198n9
Samanez-Larkin, G.R., Wagner, A.D., and Knutson, B. 196, 199n27
Samanez-Larkin, G.S., Levens, S.M., Perry, L.M., Dougherty, R.F., and Knutson, B. 185, 198n11
Sanders, J.L. and Brizzolara, M.S. 149n3
saving: homeownership and 14–16; twin investment behavior and 14–15
Schneider, M. and Prasso, S. 159–60, 167n34
Schwandt, H. 115, 119n36
seasonal affective disorder (SAD) 138–42
Segal, Nancy 26n1
Seiler, M.J. and Walden, E. 67, 77n9
self-interest 1, 159
seniors' investment decisions 191–7; *see also* cognitive aging, diminished decision making and
sensation-seeking behavior 45
Seo, M.G., Goldfarb, B., and Barrett, L.F. 128, 135n9
serotonin 64
serotonin transporter 5-HTTLPR 74
sex 42, 51
Shank, C. 167n30
Shank, C.A. 161, 168n41
Shank, C.A., Dupoyet, B.V., Durand, R.B., and Patterson, F. 158n40, 160–61
Sharpe, W. 140
Shiffrin, R.M. and Schneider, W. 64, 76n1
Shiller, Robert J. 1
Shiv, B., Loewenstein, G., Bechara, A., Damasio, H., and Damasio, A.R. 69, 77n14
Simons, D.J., Boot, W.R., Charness, N. et al. 197, 199n31
single nucleotide polymorphisms (SNPs) 34–5, 36, 38, 40
sleep, coffee and investing 95–106; Balloon Analog Risk Task (BART) 98, 102; biology and psychology in finance 6; brain activation, sleep and economic risk taking 98–9;

country-wide analysis 99–100; fragmentation of sleep 96; functional magnetic resonance imaging (fMRI) 98; global market games 101; individual focused experiments 100–101; Iowa Gambling Task 102–3, 104; memory, deep sleep and 96; Pittsburg Sleep Quality Index 97; questions 104; rapid eye movement (REM) sleep 96; reduction of sleep 96; sleep, economic risk taking and 96–8; sleep and financial decisions 99–101; sleep physiology 95–6; slow-wave sleep 96; summary 104

sleep deprivation 6; effects of 86, 95; stimulants and risk taking and 101–4; strategic thinking and 98–9

Smith, A., Lohrenz, T., King, J., Montague, P.R., and Camerer, C.F. 66–7, 77n8

Smith, P.J., Blumenthal, J.A., Hoffman, B.M. et al. 118n14

Smith, Vernon L. 1

social risk, gender differences and 42

societal differences, gender differences and 43–4

society illness, stock market performance and 112–13

Sokol-Hessner, P., Camerer, C.F., and Phelps, E.A. 70, 77n18

Sokol-Hessner, P., Hsu, M., Curley, N. G. et al. 131, 135n13

S&P 500 Index 37; returns from, influences of fine particulate matter (PM2.5) on 144

Springer, Jim ("Jim twin") 13

Stanton, S.J., Liening, S.H., and Schultheiss, O.C. 81, 94n4

Stanton, S.J., O'Dhaniel, A., Mullette-Gillman, R., McLaurin, E. et al. 81–2, 93n5

Statman, Meir 10n2

STEM (Science, Technology, Engineering, and Math) careers, underrepresentation of women in 43–4

stock markets: air pollution and 143–5; bubbles on, impact of emotion on 129; participation in 18–20; participation in, actual and perceived financial literacy in 53; participation in, financial literacy and gender for 53; participation in, IQ and rate of 170–71; price crash risk, overconfidence and 164; seasonal affective disorder (SAD)

and 139–41; stock selection 66; terrorism and 148; weather and 138

Stoll, H.R. 150n21

striatum 62

Stunkard, A.J., Faith, M.S., and Allison, K.C. 117n9

summaries: brain function, financial decisions and 76; cognitive aging, diminished decision making and 197; emotional and moody investors 134–5; environmental factors in financial decision making 148; gender differences, investments and 56; hormones, influence on financial risk taking 92; human genome, investments of adoptees and 39–40; intelligence and investment performance 180–81; sleep, coffee and investing 104; twin investment behavior 26; wellness, influence on financial decisions 116–17

Summers, B. and Duxbury, D. 127, 135n8

Sunden, A.E. and Surette, B.J. 46, 57n17

sunshine, serotonin and 137

Survey of Health, Ageing, and Retirement in Europe (SHARE): cognitive aging, diminished decision making and 192; intelligence and investment performance 172–3; wellness, influence on financial decisions 110, 114

Swedish Tax Agency 14, 15, 18, 22, 23, 32

Swedish Twin Registry: hormones, influence on financial risk taking 90; twin investment behavior 14, 15, 17, 22, 38, 90

Symeonidis, L., Daskalakis, G., and Markellos, R.N. 141, 150n24

Tanaka, H., Taira, K., Arakawa, M. et al. 95, 105n3

Taubman, Paul 14, 26n2

Tauni, M.Z., Rao, Z.-ur-R., Fang, H.-X., and Gao, M. 156–7, 167n17

Taylor, M.P. and Wozniak, D. 53, 58n35

Telephone Interview for Cognitive Status 192

temporal lobe 61–2

Ten Brinke, L., Kish, A., and Keltner, D. 162–3, 168n44

Terburg, D., Morgan, B., and Van Honk., J. 93n9

terrorism, stock market and 147–8

testosterone: biology and psychology in finance 6; economic risk taking and 81–2; financial decisions and 83–8; hormones, influence on financial risk taking 79–80; traders' profits and losses and 83

testosterone proxies 88–92; economic risk taking and 89; financial decision making and 89–91; validity of 91–2

thalamus 62–3

Thaler, R.H., Tversky, A., Kahneman, D., and Schwartz, A. 129, 135n11

Thaler, Richard 1

theory of mind 170, 178–80

Thinking, Fast and Slow (Kahneman, D.) 174

Thoma, V., White, E., Panigrahi, A., Strowger, V., and Anderson, I. 178, 182n14

Thomas, M.L., Sing, H.C., Belenky, G. et al. 105n6

Thorgeirsson, T.E., Gudbjartsson, D.F., Surakka, I. et al. 41n12

Tietjen, G.H. and Kripke, D.F. 149n6

time horizons, income needs and 191

Tobin's Q 112

Toplak, M.E., West, R.F., and Stanovich, K.E. 181n10

trading decisions, impact of gender on 48–9

Treynor, J. 140

Turcot, V., Lu, Y., Highland, H.M. et al. 41n11

23andMe 28

twin investment behavior 13–27; Add Health project 25; age, stock market participation and 20; behavioral biases 22–4; biology and psychology in finance 3; decision correlation between twin groups 19; disposition effect 23–4; econometric techniques 19–20, 21, 22; fraternal twins 13–14; gender, stock market participation and 20; genetic contribution to life satisfaction 25; genetic differences, savings behavior and 15; genetics, behavioral biases and 22–3; genetics, earnings and 14; genetics, investment choices and 24; genetics, pension choices and 21; giving 24–5; happiness 25; homeownership 15–16; identical (monozygotic) twins 13; investing 18–22; investing style 21–2;

investing style preferences, drivers of 22; mean-variance portfolio theory (Markowitz) 16; Minnesota Center for Twin and Family Research 14; Minnesota Twin Registry 16; nature, nurture, and unique experiences 13–14; nurture, pension choices and 21; nurture, savings behavior and 15; nurture experiences 19; pension choices 20–21; pension fund choices, analysis of 21; portfolio holdings, behavioral biases and 23; psychological biases 22–3, 23–4; questions 26; risk attitudes 16–18; risk aversion 18; risk investment 18; risk preferences, determination of 16–17, 17–18; risk-return models 16; saving 14–15; saving and homeownership 14–16; savings behavior, origins of 15; stock market participation 18–20; summary 26; Swedish Tax Agency 14, 15, 18, 22, 23, 32; Swedish Twin Registry 14, 15, 17, 22, 38, 90; twin contact, stock market participation and 20; twin research 13–14; unique experiences, savings behavior and 15; Value-Growth score (Morningstar) 22; value *versus* growth 21–2; wellbeing, giving, happiness and 24–5

2D:4D ratio, facial masculinity and 88–9

unique experiences: savings behavior and 15; variance decomposition attributable to 29–30

Value-Growth score (Morningstar) 22

value *versus* growth, twin investment behavior and 21–2

Van Dongen, H.P.A., Maislin, G., Mullington, J.M., and Dinges, D.F. 96, 105n10

Van Honk, J., Schutter, D.J.L.G., Hermans, E.J., and Putman, P. 82, 93n7

Vartanian, L.R. 108, 117n8

Vedel, A. and Thomsen, D.K. 167n31

Venkatraman, V., Chuah, Y.M.L., Huettel, S.A., and Chee, M.W.L. 105n7

Venkatraman, V., Huettel, S.A., Chuah, L.Y.M. Payne, J.W., and Chee, M.W. L. 98–9, 106n18

Veritas Genetics 28

voting behavior, genetics and 29

Wall Street, women on 44
Wang, A.Y. and Young, M. 148, 151n45
wealth: intergenerational transmission of 33; in retirement, educational attainment polygenic score and 36–7
weather, mood, financial decisions and 137–8
Weinstein, A.M., Voss, M.W., Prakash, R.S. et al. 118n13
wellbeing, giving, happiness and 24–5
wellness, influence on financial decisions 107–19; absenteeism 43, 111, 113; asset holdings, health and 109; Behavioral Risk Factor Surveillance System (BRFSS) surveys 111, 115; *The 100 Best Companies to Work For in America* (Levering, R., Moscowitz, M., and Katz, M.) 112; biology and psychology in finance 6–7; Centers for Disease Control and Prevention (CDC, US) 107, 111, 112–13; Dow Jones Industrial Average stock index, health impacts on 115; Dynamic Experiments for Estimating Pre-ferences survey 110–11; economic policy uncertainty, impact on healthy choices 116; employee health, company performance and 111–12; exercise and diet 108; finance, impact on health of 114–16; financial crisis (2008–2009), well-being during 115; financial distress, effects of 115–16; health and obesity 107–8; Health and Retirement Study (HRS) 109, 110, 113, 114, 115; health and risk taking, experimental evidence 110–11; Income Dynamics, Panel Study of 114; London Stock Exchange 116; mental health, financial decisions and 113–14; obesity, impact on corporate policies 111–12; obesity and health 107–8; Organization for Economic Co-operation (OECD) 107, 117n2; physical health, financial decisions and 109–13; portfolio choice, health and 109–10; questions 117; society illness, stock market performance and 112–13; summary 116–17; Survey of Health, Ageing, and Retirement in Europe (SHARE) 110, 114

Westbrook, A., Martins, B.S., Yarkoni, T., and Braver, T.S. 196, 199n28
Wisniewski, T.P. and Lambe, B.J. 116, 119n42
Worthy, D.A. and Maddox, W.T. 199n29

Xie, L., Kang, H., Xu, Q. et al. 105n14
Xu, P., Gu, R., Broster, L.S. et al. 78n19

Yogo, M. 118n18
Yoo, S.-S., Gujar, N., Hu, P., Jolesz, F. A., and Walker, M.P. 105n8

Zamarro, G. 158, 167n25
Zhang, X., Chen, X., and Zhang, X. 143, 150n32
Zhong, S., Israel, S., Xue, H., Ebstein, R.P., and Chew, S.H. 75, 78n30
Zuckerman's Sensation Seeking Scale 50
Zyphur, M.J., Narayanan, J., Arvey, R.D., and Alexander, G.J.. 16–17, 18, 26n5

Printed in the United States
by Baker & Taylor Publisher Services